Poverty Bay Books

PREVAILS

Copyright © 2021 by James R. Simpson

All rights reserved. No part of this book publication may be reproduced, stored in a retrieval system, or transmitted in any form or by any means electronic, mechanical, photocopy, recording, or any other means, except for brief quotation in reviews, without the prior permission of the author. To contact the author, email: jamesrsimpson@gmail.com.

Book design and publishing management: Bryan Tomasovich, The Publishing World
Author photo: Masaru Yamada

Simpson, James R.
AI Prevails: How to Keep Yourself and Humanity Safe

ISBN: 978-0-578-85472-4

Also for sale in paperback and ebook formats.

1. Computers / Artificial Intelligence. 2. Technology & Engineering/ Robotics.
3. Technology & Engineering/ Social Aspects. 4. Social Science/ Technology Studies

Poverty Bay Books

Distributed by Ingram

Printed in the U.S.A.

AI

PREVAILS

How to Keep Yourself and Humanity Safe

James R. Simpson

For Itsuko, Randy, Roderick, and Roberta

INTRODUCTION

1	LET THE GAME BEGIN	1
2	THE REALITY: ROBOT AND HUMANOID ROBOT NUMBERS GROWING FAST	6
3	ENDOWMENT OF ROBOTS AND HUMANOIDS WITH FEELINGS	17
4	TRANSHUMANISTS' VIEW ON A POSTHUMAN CONDITION	32
5	BENEFICIAL AND MORAL CONSEQUENCES OF ARTIFICIAL INTELLIGENCE	45
6	THE CONNECTION BETWEEN HUMANOID ROBOTS, HUMANOIDS, AND SUPERINTELLIGENCE	59
7	THE PUBLIC'S RIGHTS VERSUS THE PLUTOCRACY, TECHNOCRACY, AND GOVERNMENT	75
8	A LEGAL PROCEDURE TO PROTECT HUMANKIND	89
9	"SORRY, YOUR JOB IS LOST TO ARTIFICIAL INTELLIGENCE"	105
10	THE FUTURE OF JOBS	114
11	THE ROLE OF HUMAN RIGHTS IN PUBLIC POLICY	123
12	THE WAY FORWARD	134

BUILDING BLOCKS FOR HUMANOIDS	139
PART 1: THE PATH TO HUMANOID ROBOT AND HUMANOID DEVELOPMENT	141
PART 2: CREATING HUMANOIDS AS LIVING ORGANISMS	159
GLOSSARY	175
INDEX	191

Introduction

The world is very different now. For man holds in his mortal hands the power to abolish all forms of human poverty and all forms of human life.

—John F. Kennedy
Inaugural Address (January 20, 1961)

In this book I reveal a vision of the future that will keep you up at night. Researchers all over the world are racing to develop technologies using Artificial Intelligence (AI). Some in the scientific community are engaged in the study and development of superintelligence. If they are successful in that endeavor, the outcome would be that AI would become smarter than their creators. We have all seen sci-fi movies that imagine hellish, apocalyptic futures where AI destroys our world. But what if this nightmare scenario doesn't arrive with a boom? What if it comes on gradually—a drumbeat growing louder every day, until one day the defining roar overtakes everything? What if that day is not very far away?

While AI is deeply troubling, other sinister endeavors are also underway. Developers in robotics, the branch of technology that deals with the design, construction, operation, and application of robots, are endowing humanoids with extraordinary features. Current projections state that that within a few decades, they could be on par with or even exceed humans. If not contained, repercussions could be drastic, like the creation of a new species that could evolve from mixing humans with humanoids. Well, don't begin to worry now.

Years ago, it became evident that robots were taking jobs from blue-collar workers. I then realized these technologies were being thrust upon an unwitting public. It became clear to me that the natural trajectory of this movement was toward advanced AI. At that point, I became upset because some things are just morally wrong. That was when I decided to take on the task of writing about who, as well as what, is behind such developments. During my science-backed research as an economist, I became curious, and then passionate, about uncovering what it would take to place controls on robotics and superintelligence.

I started searching for keys about who and what is behind research and development of these technologies. Was it a cabal, which is a secret political clique or faction? Was it the technocrats, a member of a powerful technical elite or someone who advocates the supremacy of technical experts? Was it the plutocracy, formally known as government by the wealthy? In more general parlance, it is now termed any form of government in which the wealthy exercise the preponderance of power, whether it be direct or indirect. Or was it our government? My conclusion was that the answer is a loose grouping of the three. While not a cabal, and not coordinated, collectively this grouping has the power to decide about technological matters. In our case, it is control of AI and robotic innovations. The conundrum highlighted in this book is whether our government could break away from the other two parts in order to successfully carry out indispensable regulations on those two technologies.

Join me as we unlock the secrets of legal procedures, the heart in a course of action. To that end, this treatise seeks to demonstrate that the most potent and feasible solution is adoption of the internationally used Precautionary Principle in courts to litigate controls and bans on technologies that have economic and moral consequences. This legal method is focused on social responsibility to protect the public from harm when scientific investigation has found a plausible risk. In our situation, the concern is scientists and others who believe they have unfettered *freedom to* develop any technologies they desire regardless of consequences to humankind.

Is it possible we can control intelligent machines, or will they control us? What would it mean to be human in an age of AI when intelligent machines coexist with or replace us? Are those around the world who develop technologies that will dramatically affect our destinies truly concerned about life satisfaction, quality of life, or whatever else happiness is termed? Should we simply internalize a life of uncontrolled artificial intelligence and bend to prophet-like futurists' scenarios? Critically, a plausible risk is that unenhanced humans could become superfluous as soon as the 2050s.

1

LET THE GAME BEGIN

We live in a society exquisitely dependent on science and technology, in which hardly anyone knows anything about science and technology. This is a clear prescription for disaster.

—Carl Sagan, *The Skeptical Inquirer* (Volume 14-3, Spring 1990)

What you are about to read is not science fiction; it is the reality of a dizzying warp-speed race to technologize America, its citizens, and ultimately the world as we know it. Do you really want to be entertained? In a grim sort of way? As expected, there are complications. Technology developers might make it possible to build a machine more intelligent than any human. If the enhanced machine breaks away from its creators, it could then potentially even write its own source code to become more intelligent than humans. But isn't that an inhuman thing to do?

Critical questions are in order.

Are technology developers around the world truly concerned about saving humankind as we know it? What would it mean to be human in an age of AI when intelligent machines coexist with or replace us? Should we citizens simply internalize a life of uncontrolled artificial intelligence? Should we bend to prophet-like futurists' scenarios of doom and gloom?

Will we control intelligent machines, or will they control us?

The issue in this book is that, although the development of superintelligence and humanoids that exceed humans' powers may not entail the extinction of all intelligent life, those two technologies could lead to permanent destruction of a great part of humanity's potential. Why would anyone want to bring on such a calamity? Could such an event truly happen in our lifetimes?

Will developers of superintelligence and robotics control us, or will we control them?

Superintelligence is any intellect that greatly exceeds the cognitive performance of humans in virtually all domains of interest. Is it possible to believe that machines (as they become smarter) and/or developers or rogues around the world with malicious or radical beliefs will always have the world's wellbeing as their primary goal? Would you trust use of a Friendly AI concept in which humans and machines would be created in a simultaneous fashion, with each depending on the friendliness of the other?

Existential risk is a risk that cannot be undone, one that poses permanent, large, negative consequences to humanity. *Singularity*, the theoretical emergence of superintelligence through technological means that could happen suddenly, within the next few decades, embodies that condition. The future that would result would be one in which some would no longer be humans by our current standards. A great part of humanity's potential would be permanently destroyed. Only a tiny fraction of all humans would get to enjoy the benefits (if any) of a posthumanity condition.

The question of gaining happiness from any kind of artificial intelligence is an intriguing one. The term *transhumanism*, coined by the biologist Julian Huxley (the brother of Aldus Huxley, author of *Brave New World*) in 1957 was "man remaining man, but transcending himself, by realizing new possibilities of and for his human nature."[1] The World Transhumanist Association (WTA), founded in 1998, focused on recognition of transhumanism as a legitimate subject of scientific inquiry and public policy. Transhumanists emphasize that although humans and individuals matter, if rational thinking and rational means are promoted, the human

---

1 Nick Bostrom, "A History of Transhumanist Thought," 2005. http://www.nickbostrom.com/papers/history.pdf. See also, Wikipedia, s.v. "Transhumanism," downloaded May 21, 2014, http://en.wikipedia.org/wiki/Transhumanism.

organism can be improved. They argue that technological means can be used beyond traditional humanistic methods to eventually enable humans to move beyond what some would think of as "human."[2]

Singularitarianism, akin to transhumanism, is the belief that technological creation of smarter-than-human intelligence will be a watershed moment in history, perhaps more comparable to the rise of *Homo sapiens* than to past breakthroughs in technology.[3] To adherents of this philosophy, the prospect of superintelligence and technological singularity is not scary—it is worth embracing. It is simply a leap to a different and better quality of life in a posthuman condition. But some of those individuals would no longer be humans by our current standards. So, could, would, those humans be happier than when they were not enhanced or were mildly enhanced humans? Would their lives be more meaningful and enjoyed than if they were not enhanced?

THE POWERFUL SEEK TO CONTROL OUR FUTURE

Should the populace be content to relegate our future and happiness to the whims of those in positions of power about regulations on superintelligence? How about humanoids? These creatures are prognosticated to be on par with humans by the mid to later 2030s. If bans are not placed on them, radically enhanced ones could feasibly make choices that might result in a new species' being created.

If the populace is concerned, should citizens fight back, and if so, how? In fact, as citizens, do we even genuinely want to be in control of our lives? Let's face it: citizen control will be difficult. The height of arrogance in America, and in some other countries, is the belief by the elite that ordinary citizens cannot be trusted to make properly informed choices about weighty matters, such as their own and the world's future.

At this point, trust in government to do the right thing and

---

2   Taken from Bostrom's Version 2.1, *The Transhumanist FAQ: A General Introduction*.

3   Paraphrased from page 44, WTA [World Transhumanist Association], *The Transhumanist FAQ*.

stand up to the elites is so low that the triumvirates' power has a clear advantage. In the past, the term *triumvirate* referred to a group of three men responsible for public administration or civil authority. In the present, it refers to a group of people representing three instruments of power: the technocracy, the plutocracy, and our national-level government. Unfortunately, the triumvirate has time on their side to advance technologies if no strong opposition emerges. So, in reality, what can be done by the citizens to assure appropriate measures are taken, at prudent, well-judged times, to avoid catastrophes resulting from superintelligence and radical humanoid development?

A day of reckoning will come faster than we can imagine. As an inducement to mull over that fact, consider the well-known researcher Ray Kurzweil's thesis that singularity should be advanced as soon as possible because "waking up the universe, and then intelligently deciding its fate by infusing it with our human intelligence in its nonbiological form, is our destiny."[4] Is this a scenario you find attractive?

Fight for Human Destiny

Who is, or should be, in charge of guaranteeing the right to control our destiny, in effect our right to happiness as stated in the Declaration of Independence? To me, without question, citizens should be in control of decision-making. But Congress and the administration are so entrenched in political infighting that we are becoming prisoners in what is taking on the trappings of a dystopia. I don't know about others, but I contend that at a bare minimum, our government must live up to the ideals we cherish in our republican democracy. That means government of the people, by the people, for the people.

But difficulties abound in these perilous times. It will take courage on the part of Congress, the administration, and the public to convince technocrats and plutocrats that they do not have *freedom to* do anything they please regarding technological developments. The reality is that a sort of Armageddon will result if regulations, controls, or bans are not placed on these two technologies soon.

---

4   Ray Kurzweil, *How to Create a Mind: The Secret of Human Thought Revealed* (New York: Viking Penguin, 2012), 282.

I will disclose things you don't want to know. A child born in 2010 will be twenty-five years old in 2035. That is when advanced humanoids with truly humanlike features and characteristics begin to fill positions once taken by humans. Does that bother you? Five years later, in 2040, your child will be thirty. By then, significantly advanced humanoids as living creatures are prognosticated to be on par with humans. Does that scare the wits out of you? Why not? Sometime in the following five years, some humanoids will have capacities that exceed those of humans if there are no regulations or bans. Terrified now? There is a way to prevent your child from going through this dreadful situation.

2

THE REALITY: ROBOT AND HUMANOID ROBOT NUMBERS GROWING FAST

Technology…is a queer thing. It brings you great gifts with one hand, and stabs you in the back with the other.

—C.P. Snow, as quoted in the *New York Times* (March 15, 1971)

Projections suggest that a recession is in store beginning in the early 2020s. So, let's work together on the twenty-dollar question of whether good times for robotics development, sales, and use will continue. What would be the effect of various levels of recession on creations for industrial use and for service robots in white-collar workplaces? Humanoid robots and humanoids are flexible in numerous white-collar job settings. So, would businesses increasingly adopt robotic innovations to reduce costs as they struggle in a changing landscape? Would AI innovations and research on robots and humanoids keep chugging right along?

ROBOT AND HUMANOID ROBOT CREATION TO THE PRESENT

Robotics is the branch of artificial intelligence that deals with robots. Let's pause for a moment to define a few terms to help visualize the role of robots, humanoid robots, and humanoids in our not-too-distant future. There is no consensus on what constitutes a robot, even in the field of robotics. A general description is a mechanical or artificial device guided primarily by a computer program or some electronic method. The easiest explanation has been *you know one when you see one*. At least you could—until a few years ago. Now, in our supercharged technological world, the term *robot* includes all sorts of programs and mechanical and digital devices.

Popularly, the terms *humanoid robots* and *humanoids* share the same definition: a robot based on the general structure of a human that has an appearance resembling that of a human being. Confusingly, an *android*, widely written about and starring in movies and TV programs, is also a humanoid robot designed to look and act as much as possible like a real person. Equally bewildering, the terms apply to both sexes, although technically, androids are the male form and females are *gynoids* (a.k.a. *fembots*). The terms *bio-android* and *android* are interchangeable. *Droid*, a robot in science fiction, is an abridgement of both android terms.

The use of the term *humanoid*, rather than *android* or *humanoid robot*, has been an attempt to humanize the mechanical being and make it more acceptable, even lovable, in daily life. The stumbling block is that there is no black-and-white distinction between a humanoid robot and a humanoid. However, that difference will become increasingly perceivable as humanoid robots gain humanlike features, begin to mix with humans, and have rudimentary professional jobs. *Humanoid* is the term used apart from robots in this book as a catchall for humanoid robot and humanoid, unless otherwise denoted. At some point, you will know a true humanoid when you see one.

Somehow, the notion of humanoid robots and humanoids as part of home life among the general populace strains the imagination, despite the new reality of the rush to buy smart devices and robotic toys. However, in Japan at SoftBank, robots are family. SoftBank has updated Pepper into a genuine humanoid robot companion that communicates "in an intuitive way, through its body movements and its voice." It can be individualized to offer content by downloading software applications that, based on your voice, the expression on your face, your body movements, and the words you use, can interpret your mood and emotions.[5]

---

5   "Pepper, the humanoid robot," AdAstra, 2020, https://sk.adastragrp.com/en/healthcare/pepper/#:~:text=Pepperismuchmorethan,adaptsitsbehaviortothem.

AI PREVAILS

An example of how AI and robotics is engaging the whole family and the workplace is AVA, a 2018 *chatbot*, which is a computer program that simulates human conversation through voice commands or text chats or both. Paradoxically, the COVID-19 epidemic has resulted in numerous technical and social innovations, one of which is that closed offices keeping workers locked up at home has resulted in their turning to PayPal for their purchases. The explosion in demand has led to the use of chatbots for communication; the medium had reached a record 65% of message-based customer inquiries by late April of 2020.

Additionally, there are *cobots*, or co-robots (from collaborative robots) that are intended to physically interact with humans in a shared workspace. Offices, college campuses, and other locations in numerous countries have leaped, in a matter of just two years, to using food delivery services by robots. In a new twist, a Chinese girl bought a robot that mimicked her handwriting, enabling her to avoid the tedious repetitive homework involved in learning Chinese characters.[6]

AI does provide benefits. However, there are social drawbacks, for "a new wave of automation could also mean that when companies start hiring again, they do so in smaller numbers. 'This may be one of those situations when automation does substantially depress rehiring,'" said Mark Muro, a senior fellow at the Brookings Institution who studies labor markets.[7] Companies and other enterprises that seek to reduce costs will increasingly engage humanoids.

Let's shift gears again. The public is aware of and informed about fictional robots and androids. However, few people realize that the origins of today's robots date back over two thousand years, during which time only slight modifications have been made.[8] These fictionalized automatons are a regular occurrence in the works of Homer, Plato, and other classical

---

6   Daniel Victor and Tiffany May, "Chinese Girl Finds a Way Out of Tedious Homework: Make a Robot Do It," *New York Times*, February 21, 2019.

7   Michael Corkery and David Gelles, "Robots Welcome to Take Over, as Pandemic Accelerates Automation," *New York Times*, April 10, 2020, https://www.nytimes.com/2020/04/10/business/coronavirus-workplace-automation.html.

8   The list of fictional robots and androids is huge, occupying 30 printed pages in Wikipedia: see "List of fictional robots and androids," accessed November 4, 2020, https://en.wikipedia.org/wiki/List_of_fictional_robots_and_androids.

authors. *The Leizi* described the first automaton in the vicinity of 250 BCE. Around 50 CE, the Greek mathematician Hero of Alexandra described a machine that automatically pours wine for party guests. In 1206, Al-Jazari described a band made up of humanoid automata that, according to Charles B. Fowler, performed "more than fifty facial and body actions during each musical selection."[9] An elephant clock that incorporated an automatic humanoid robot mahout (a person who works with, rides, and tends an elephant) struck a cymbal on the half-hour. Leonardo da Vinci designed a humanoid automaton (a self-operated machine) that looked like an armored knight in 1495. About 100 years ago, Karel Čapek was credited with coining the term *robot* in *R.U.R. (Rossum's Universal Robots)* in 1921.

After that, there came an exponential explosion of interest in robots that continues unabated in its growth today. The first robot put to useful work was named Televox and was created by Westinghouse in 1926. The Westinghouse Corporation then created a humanoid robot known as Elektro in the 1930s. Elektro was on exhibition in the 1939 and 1940 World's Fairs. In the 1950s, George Devol and Joseph F. Engelberger started Unimation, the world's first robot manufacturing company. The first industrial robot was Ultimate, which worked on a General Motors assembly line in 1961.

Robots and androids have spawned a gold rush in the form of books, movies, and television shows thanks to people's intense fascination with these fictional and real creatures. Fritz Lang's film *Metropolis* depicted gynoid robots called Parody, Futura, and Robotrix in 1927. In the 1930s and 1940s, Isaac Asimov published short stories on robots. His first novel, *Pebble in the Sky*, was published in 1950. His short story collection called *I, Robot* came out the same year. The *Foundation* series was first published in 1951 and continued on into the 1980s. Jack Williamson picked up on that theme and published *The Humanoids* in two novels, published in 1949 and 1980 respectively.

Science fiction (a.k.a. sci-fi) took off in 1950 with *Ro-man*, a robot bent on destroying the earth. There were even radio series in the 1970s and 1980s that featured robots, such as

---

9   Charles B. Fowler, "The Museum of Music: A History of Mechanical Instruments," *Music Educators Journal* 54, no. 2 (October 1967): 45-49.

Marvin the Paranoid Android in *The Hitchhiker's Guide to the Galaxy* BBC radio series (1978–1980). Tidy, George, Fagor, Surgeon General Kraken, and miscellaneous other androids were featured in James Follett's *Earthsearch* BBC radio series that ran 1980–1981. "Accident-prone and apologetic gopher robots" were featured on the BBC radio series *Nineteen Ninety-Four* in 1985. *The Digital Human* on BBC Radio had its last episode in March 2018.[10]

The film *The Clever Dummy* played in 1917, when the term "robot" did not yet exist. Television films and series began with *The Adventures of Superman* (1952–1958) and "The Runaway Robot" episode (1953).[11] The number of comic books, graphic novels, comic strips, web series, and video games is impressive and growing, thanks to animation.

It has taken a relatively long time for robots to globally make significant inroads in industry, the service sector, and agriculture. There were just 55,000 industry robots worldwide in 2002. That was the beginning of a torrent of innovation and realization about the value of robots. The number leaped to 2.4 million in 2018, and 3.8 million units are projected for 2021. Continuous rapid growth is forecast.[12]

Development of Humanoid Robots and Humanoids in the Modern Era

Creation of humanoids, as we now know them, that have humanlike features, began on a robot platform. Waseda University in Japan initiated the WABOT project in 1967, completing it in 1972 with WABOT-1, the world's first full-scale, humanoid, intelligent robot. It took Waseda University twelve years to finish Wabot-2 in 1984. The outcome was stunning: a humanoid robot musician able to "converse with a person, read a normal musical score with its 'eye' and play

---

10 Wikipedia, s.v. "List of fictional robots and androids," accessed November 4, 2020, https://en.wikipedia.org/wiki/List_of_fictional_robots_and_androids.

11 Ibid.

12 The International Federation of Robotics, 2018, https://ifr.org/downloads/press2018/Executive_Summary_WR_2018_Industrial_Robots.pdf.

tunes of average difficulty on an electronic organ."[13] Six years later, in 1990, Tad McGeer showed that a biped mechanical structure with knees could walk passively down a sloping surface. Waseda University then developed, in 1997, Hadaly-2, a humanoid robot that interacted with humans both verbally and physically.

Development of humanoids began to accelerate at an exponential rate in the year 2000, when Honda created its eleventh Bipedal Humanoid Robot that was able to run. In 2005, Mitsubishi Heavy Industries released Wakamaru, a Japanese domestic robot "primarily intended to provide companionship to the elderly and people with disabilities."[14] NASA and General Motors finalized Robonaut 2, a very advanced humanoid robot, in 2010. That robot had the distinction of being part of the payload in the successful launch of space shuttle *Discovery* on February 24, 2011.

Singapore's Nanyang Technological University released Nadine in 2015. This socially intelligent humanoid robot is a realistic-looking gynoid who is friendly and greets you in response to your greeting. She makes eye contact and remembers all of the conversations you've had with her. That same year, Hanson Robotics in Hong Kong released a humanoid robot named Sophia, designed to look like Audrey Hepburn. That creation was updated as a social robot in 2018. Since then, much more has taken place in the robotics field that is genuinely startling.

PROGNOSTICATIONS ON HUMANOID ROBOT AND HUMANOID DEVELOPMENT

I have divided humanoid development into three stages that merge into each other: humanoid robot, advanced, and radically enhanced. To gain a reasonable idea of what may come or is likely to come in the future, visualize a child born in 2015 as he or she grows up (Table 1):

---

13 "Robot Plays the Organ, Next He Will Rip Your Arms Off," Synthtopia, October 29, 2000, https://www.synthtopia.com/content/2009/10/29/robot-plays/.

14 "Robot news," Robotnews, April 6, 2007, https://robotnews.wordpress.com/2007/04/06/wakamaru/.

Table 1: Timelines of humanoid development compared with the age of a child born in 2015.

2015		Child born.
2025	10	First humanoid robots begin to mingle with humans.
2030	15	Moderately augmented humanoid robots begin to mix with humans and hold rudimentary professional jobs.
2035	20	Early advanced humanoids with truly humanlike features and characteristics begin to fill positions once taken by humans.
2040	25	Significantly advanced humanoids as living creatures are on par with humans.
2045	30	A period of radical enhancement takes place around 2040–2045, during which some humanoids have capacities exceeding those of humans.

Now a conundrum. The term *radically enhanced humanoids* refer to those in an extremely advanced stage of development or beyond. They have such advanced capacities that they exceed those of present humans. If there are no bans or controls on humanoid development by the time these living creatures are on par with humans, at that time some would no longer be unambiguously humanlike by our current standards. Some, particularly if programmed with free will, might have the capacity to make choices that would adversely impact humankind to such an extent that a new species might be created. The result would be a posthumanoid condition.

There is, however, an alternative to these dire scenarios. If people have discerned by the mid-1930s there is an existential risk (a risk that cannot be undone and that poses permanent, large, negative consequences to humanity) there would be sufficient time to begin monitoring humanoid development. Then strict controls and bans would have to be placed on further development before 2040, when living creatures are on par with humans. A child born in 2015 would be 25 years old at that time.

It is crucial to understand that the global nature of robotics activity makes enforcing bans and controls very difficult. Now Further, dates in the prognostications are not fixed. However, they are not speculations. Neither are they predictions of what will happen, because the events depend on economic conditions, the interest of robotics creators, their funding, profits to developers, and societal acceptance of service robots and humanoids in white-collar jobs. A natural question is on what timeline are these prognostications based. A succinct answer is that I have extensive references from my in-depth research beginning a half-dozen years ago.

It is important to understand that humanoid development is not just on the drawing board; rather, it is in the humanoid robot stage. Many of the techniques and technologies are already well advanced beyond the current attributes now found in robot humanoids. Detailed information and projections of times when developers might imbue humanoids with further attributes designed to make them look and act as much like a real person as possible are found in the section, "Building Blocks for Humanoids."

Part 1. Path to Humanoid Robot and Humanoid Development
- Close Coupling of Humans and Machines
- Mobility, Vision, and Appearance of Robots
- Body Parts Regeneration via 3-D Printing
- Speech and Power to Operate
- Transplants of Body Parts
- Reproduction: Womb Transfer and Surrogate Babies
- Embryo Transfer

Part 2. Creating Humanoids as Living Organisms
- Some Genomics Are Truly Unnerving
- Cloning
- Cloning Humans
- Thinking, Emotions and Brains for Being a Living Organism
- Should Humanoids Be Given Free Will?
- Humanoids as a New Species

Also helpful is the glossary, which contains 150 definitions of techniques, technologies, and other supporting material.

Human and Robotic Workforce

The number of researchers who are directly or indirectly involved in technology development worldwide that is related in some way to robots and humanoids is overwhelming. In brief, thousands of individual and collaborative activities are taking place in numerous scientific fields that contribute directly or indirectly to humanoid development. For example, in 2017 the Microsoft Academic Search listed 11,338 organizations for computer science and, if cognate disciplines are also counted as artificial intelligence, the number of AI organizations was 21,802 in 2014.[15]

From 2012 to 2016, U.S. science and engineering employment increased at a compound annual growth rate of 2.9 percent, while overall employment grew by 1.9 percent. Literally hundreds of thousands of scientists just in America, and hundreds of thousands more worldwide, are engaged in some aspect of fields related in some way to robot, humanoid, and superintelligence development, even if peripherally (*superintelligence* is "any intellect that greatly exceeds the cognitive performance of humans in virtually all domains of interest").[16] The Congressional Research Service alone listed 6.9 million people in the U.S. as scientists and engineers in 2016, accounting for 4.9 percent of total U.S. employment.[17] At that time, people were engaged in these fields in the following percentages:

- computer-related occupations: 58%
- engineers: 24%
- science and engineering managers: 8%
- physical scientists: 4%
- life scientists: 4%
- mathematics occupations: 2%

---

15   Luke Muehlhauser, "How Big is the Field of Artificial Intelligence? (initial findings)," MIRI Machine Intelligence Research Institute, January 28, 2014, https://intelligence.org/2014/01/28/how-big-is-ai/.

16   Nick Bostrom, *Superintelligence: Paths, Dangers, Strategies* (New York: Oxford University Press, 2014), 22.

17   John F. Sargent Jr., "The U.S. Science and Engineering Workforce: Recent, Current, and Projected Employment, Wages, and Unemployment," Congressional Research Service, November 2, 2017, https://fas.org/sgp/crs/misc/R43061.pdf.

The International Federation of Robotics (IFR) reported in 2018 that worldwide operational stocks of industrial robots were as follows:[18]
2012—1.2 million
2016—1.8 million
2018—2.4 million

And projected as follows:
2019—2.8 million
2020—3.2 million
2021—3.8 million, a compound annual growth rate of 14 percent from 2012 to 2021, and 17% overall from 2019 to 2021.

The IFR also reported on the total number of professional service robots, with occupations in fields that include public relations, medical operations, and assistance to the elderly and people with disabilities. This grouping rose 85 percent between 2016 and 2017. The projections are for 21 percent annual growth through 2021. Prognostications of growth farther out for robots, humanoid robots, and true humanoids are hard to make, as are those for other AI innovations. The best one can say is that growth will increase "a lot." However, one thing is for sure: the invasion of industrial robots will escalate, and service robots will grow early on at an exponential pace. The reason is I am confident that scientists and developers will be intrigued by the challenges because of their great, almost overpowering, desire to create and their love for creating new things.

Finally, be aware that I do not attempt to provide a complete diagnostic of all the technological aspects in the creation of humanoids in this book, because so many diverse fields are involved and new material becomes available every day. What I *do* provide are my well-reasoned explanations and conclusions grounded in my economic and scientific background.

The social side of robotics is not only fascinating—it is reality unfolding. Some people fall in love by exchanging letters or through cyberspace romances; others prefer the companionship of live or stuffed animal robots to people. Can the wave of alternative new lifestyles lead to a common phenomenon of having a humanoid companion as a principal

---
18 Ibid.

ingredient in one's life? Let's dig deep. Some robotics experts predict that reasonably soon, we'll be making love to sexbots with romantic feelings and marriage in mind instead of just having sexual intercourse with them.

3

Endowment of Robots and Humanoids with Feelings

Science is the first of sins, the germ of all sins, the original sin. This is all there is of morality—"Thou shalt not know"—the rest follows from that.

—Friedrich Nietzsche, *The Antichrist* (1895)

Can you imagine two office workers, one a human, the other a humanoid, working closely together on various tasks and projects? That over time, a strong bond forms as they become familiar with one another's idiosyncrasies and dependent on each other's strengths and attributes? They come to prefer their working relationship above all others and become inseparable, both on the job and away from work. Would it be reasonable to assume that, as their relationship develops, they might want to celebrate their happiness and harmony together through a committed, long-term partnership? True, humanoids are not warm-blooded mammals. However, think about this moral philosophical issue: if humanoids can profess love for their partner, and the recipient accepts that love, then what is wrong with that? In today's world, where a wide diversity of partnerships has become accepted, why not legislate that when humans and humanoids fall in love, it would be legal for them to marry?

Animaloids as Trusted Companions to Alleviate Loneliness

One advance in robotic developments inexorably leads to another. The law of markets says that supply creates its own demand. That law is literally true in the AI race to develop new technologies, many of which will become essential in man–computer symbiotic development of humanoids. One such technology is algorithms used to discern humans' preferences for enhanced interaction with a humanoid. Sensors have been fashioned to help humanoids read a person's facial expression and thus that individual's feelings. Advances are underway in cognitive architectures to enable a robot or humanoid to understand how it may better motivate a human.

Social robots hold great promise for helping an aging population, from offering reminders to take medicine to assisting in staying in better touch with friends and family. James Young, a researcher at the University of Manitoba, points out that if robots are to catch on across all ages, they need to prove themselves useful and helpful. "Whether that's by helping with loneliness, helping with tasks like cooking, that's the key," he said. "Once people are convinced something is useful or actually saves time, they're really good at adapting."[19] That includes the wide scope of intelligent machines—robots—that can help and assist humans in their day-to-day lives. An example is a seal-looking animaloid named Paro. That robot was created as a synthetic organism designed to look and act as much as possible like a real animal. It has been introduced into workplaces for therapy with aged or disabled people, and it has been a success with patients.[20]

Societal mores are changing, and the issue has become who or what should call the shots on social robots. Early on, the issue was whether the decisions should rest with ethicists, who have their own feelings and values, or whether the task could also fall on relatives, patients themselves, caregivers, and developers of robotics devices. Those developments sound benign. But it turns out there are ethical problems, because this animaloid *does* look and act as much as possible like a

---

19   Matt O'Brien, "Robots are getting more social. Are humans ready?" *Seattle Times*, August 13, 2018, https://article.wn.com/view/2018/08/13/Robots_are_getting_more_social_Are_humans_ready/?section=More+News&template=worldnews%2Findex_backup.txt.

20   Jennifer Levitz, "Communities Struggle to Care for Elderly, Alone at Home," *Wall Street Journal*, September 25, 2015, http://www.wsj.com/articles/communities-struggle-to-care-for-elderly-alone-at-home-1443193481.

real animal, and some patients may think Paros are real, living friends. Research has provoked debate about the extent to which a humanoid robot should look humanlike, because that sort of anthropomorphizing may leave a patient feeling unease or even disgust. Consequently, some argue that that it is best to keep social robots machinelike. Others believe that imbuing lifelike humanoid robots with realistic human or animal features risks deception. Still others argue that having lifelike robots and humanoids in the company of children or the aged will engender love and compassion.

Margaret Rouse posted about the advantages and disadvantages of social robots.[21] "While social robots employ leading edge technology, they're not humans and lack empathy, emotion and reasoning. They handle routine tasks that they are programmed to do, but may respond unpredictably to situations for which they were not trained. As with any technology, robots are susceptible to hardware malfunctions and failures and may involve a high cost to repair and maintain. In addition, humans that develop an over-dependence on social robots, such as for emotional companionship, may miss out on person-to-person interactions that are the essence of the human condition." Meanwhile, the scientific community marches on with grand plans to enable robots, humanoid robots, and even dogs to carry out tasks assigned to them.

On an immediate and practical side, an increasingly large proportion of the population has pets. Dogs are ubiquitous. You see them led around in grocery stores. They are parked curbside at restaurants. They hang out car windows. They are brought on airplanes as caregivers. One 2015 headline read: "Haute dogs fill social calendars—and closets—for Halloween."[22] The newspaper article also revealed that 20 million pet owners are planning on spending about $350 million dollars on costumes for their pets this Halloween. That's not all. A significant problem with pets is that live

---

21   TechTarget Network, "Definition: Social robot," Whatis.com, accessed November 4, 2020, https://searchenterpriseai.techtarget.com/definition/social-robot.

22   Sue Manning, "Haute dogs fill social calendars—and closets—for Halloween," *Seattle Times*, October 29, 2015, http://www.seattletimes.com/nation-world/haute-dogs-fill-social-calendars-and-closets-for-halloween/.

animals do require an extraordinary amount of effort.[23] They are problematic for harried individuals who are too busy to walk, feed, or clean up after a living creature. Economics is another consideration. Expenses on animals run upward of $50 billion annually in America for veterinary care and food.[24] Nevertheless, demand for dogs as companions surged during the COVID-19 pandemic.

Abuse of animals is also an increasing concern. By 2018, bills on animal abuse had been introduced in eleven state legislatures. An equal number more had been considering them. Then, on November 25, 2019, President Trump signed a law that made cruelty to animals a federal crime. According to NPR, "The penalty for violating the law can include a fine, a prison term of up to seven years, or both."[25] The bottom line is that many people just feel longing for some kind of companionship, and they don't want their companions to be abused, whether they're humans, other creatures, or a robotic innovation.

Well, wouldn't some consider substituting an animaloid for a live animal? The answer is yes. Animaloids are a wave of the future. Aibo, the robot dog made by Sony that went on sale in September 2018, is a harbinger of affectionate robots that will become autonomous companions, like members of the family. As Geoffrey A. Fowler described, Aibo is about the size of a Yorkshire terrier, has twenty-two joints and lifelike movements, and can lie down and flip over to play dead upon hearing the words "bang-bang." Cameras are built into its nose and lower back to help it wander around the house like a Roomba. It also has four microphones that let it hear commands and figure out who's issuing them, is always on line, and it doesn't require

---

23   We have clearly entered into an era that in some ways is characterized by love in the time of pets. For example, Ellen Byron, "The Joy of Cooking for Dogs," *Wall Street Journal*, May 28, 2014, http://online.wsj.com/public/resources/documents/print/WSJ_-D001-20140528.pdf.

24   There is the money side. The public spent $55.7 billion on pets in 2013, as reported by Sue Manning, "We lavish attention on pets, products industry benefits," *Seattle Times*, March 14, 2014, http://seattletimes.com/html/businesstechnology/2023125076_petsspendingxml.html.

25   Richard Gonzales, "Trump Signs Law Making Cruelty to Animals a Federal Crime," NPR, November 25, 2019, https://www.npr.org/2019/11/25/782842651/trump-signs-law-making-cruelty-to-animals-a-federal-crime.

apps, among other features.[26] This animaloid, priced at $1,800 in 2020, is a work in progress whose price will fall dramatically, as have those of other smart thing releases. However, even at that price, the economics are on the positive side when compared to the average lifetime cost of a real dog, calculated in Credit.Com in January 2019 at somewhere between $1,290 and $6,445.

As Fowler explains, here's why Aibo matters, despite its current limitations. "I fell for it. Over two weeks of robot foster parenting, almost every person I introduced Aibo to went a little gaga. ... Affectionate robots have the potential to comfort, teach, and connect us to new experiences—as well as manipulate us in ways we've not quite encountered before." That pronouncement and other data explain why animaloids can be expected to expand in use.

Robotic personal assistant social robots abound. Embodied Intelligence and other companies are now using the latest advances in deep reinforcement learning and deep imitation learning to develop teachable robots. The latest is *few-shot learning*, the practice of feeding a learning model with a very small amount of training data. The approach is to use repetition, as people do when learning which particular movements bring success and which do not.[27] The idea, according to Evan Ackerman, is that "with a flexible enough learning framework, programming becomes trivial because the robot can rapidly teach itself new skills with just a little bit of human demonstration at the beginning."[28] Thus, it's not hard to visualize using transfer learning—the ability to use knowledge previously gained from one context in another context—to teach low-cost robots to perform equally well. Teaching humanoids to do office work in which they interact with humans and with other robotic personal assistants is along the same line. But it requires considerably more effort and creation of methods. Think about upgraded chatbots.

26   Geoffrey A. Fowler, "Aibo the robot dog will melt your heart with mechanical precision," *Seattle Times,* September 28, 2018, https://www.seattletimes.com/business/aibo-the-robot-dog-will-melt-your-heart-with-mechanical-precision/.

27   Evan Ackerman, "AI Startup Embodied Intelligence Want Robots to Learn From Humans in Virtual Reality," Spectrum.ieee.org, https://spectrum.ieee.org/automaton/robotics/artificial-intelligence/ai-startup-embodied-intelligence, November 8, 2017.

28   Ibid.

Robots, Humanoid Robots, and Humanoids

The animaloid and dog examples are just the tip of the iceberg. Issues surrounding humanoid consciousness are weighty and realistic, considering the inevitable advances that AI scientists envision for the next few decades. That is why acclaimed author David Levy, in 2006, placed such great emphasis on the likelihood of robots' having consciousness.[29] If they had consciousness, they would hopes, wishes, beliefs, dreams, and feelings. The topic is a profound one, as is expressed in the 2017 book *Robot Sex: Social and Ethical Implications*, published by MIT Press, in a chapter by co-editor John Danaher.[30] In fact, sexbots have gained widespread acceptance as conveyances for personal sexual satisfaction and as sex workers in commercial establishments such as brothels.[31] Thanks to the more humanlike skin texture they now have, they are already standard fare in parts of Europe and Japan.[32]

The wheels of innovation are turning so rapidly that in 2017, the world's first robo-bordello—Barcelona's Lumidolls—promised customers they would "struggle to distinguish between the lifelike dolls and the real thing."[33] By 2018, KinkySdollS, a Canadian company with one brothel in Toronto, planned to open a "love dolls brothel" in Houston.[34] The crux of the matter is that, as ludicrous as it may seem, some researchers have concluded that robot humanoids do

---

29 David Levy, *Robots Unlimited: Life in a Virtual Age* (Wellesley, Massachusetts: A K Peters, Ltd., 2006).

30 John Danaher and Neil McArthur, eds., *Robot Sex: Social and Ethical Implications* (Cambridge, Massachusetts: Massachusetts Institute of Technology, 2017).

31 See, for example, Peter Nowak, *Sex, Bombs and Robots: How War, Porn, and Fast Food Shaped Technology as We Know It* (Guilford, CT, USA: Lyons Press, 2011), http://www.sexrobot.com/peter-nowak-sex-doll-brothel-sex-robot.

32 Shirley S. Wang, "Closing in on the Formula for Artificial Skin," *Wall Street Journal*, July 6, 2010, http://online.wsj.com/news/articles/SB10001424052748704293604575343033962283238.

33 Paul Harper, *The Sun*, October 27, 2017, https://www.thesun.co.uk/news/4779261/inside-the-barcelona-brothel-where-women-have-been-replaced-by-inflatable-dolls-so-randy-punters-can-fulfil-fantasies-they-wouldnt-dream-of-revealing-to-a-human/.

34 Juan A. Lozano, "Mayor, others push back on proposed robot brothel in Houston," AP News, September 27, 2018, https://apnews.com/8171bff07302452d9b3d-2d5acc7214fa.

have feelings and thus can be concerned about harm to them or about harming others. That aspect of AI is not new. Laws of robotics and ethics have been preparing for these issues since way back in the WWII era.

Asimov's Three Laws of Robotics

Developing the ability of robots, and by extension humanoids, to defend themselves and not harm others is one of the crucial steps in the three stages of humanoid development outlined by Isaac Asimov in 1942. Currently, robots and humanoids do not have outward feelings that would allow us to classify them as a living entity. However, technological development is closing the gap between the rules specified by Asimov and the forthcoming specter of humanoids imbued with life. The Three Laws of Robotics in Asimov's 1942 short story "Runaround." The Three Laws, quoted as being from the *Handbook of Robotics, 56th Edition, 2058 A.D.*, are:

> First Law
> A robot may not injure a human being or, through inaction, allow a human being to come to harm.
>
> Second Law
> A robot must obey the orders given it by human beings, except where such orders would conflict with the First Law.
>
> Third Law
> A robot may not injure one of its own kind and it must defend its own kind unless that robot is interfering with the first or second rule.[35]

Advances in technological innovations provide evidence that other ethical rules are needed beyond those three rules. Take, for example, the desirability of a robot's ability to defend

---

35   Isaac Asimov, "Runaround," in I, Robot (The Isaac Asimov Collection) (New York: Doubleday, 1988). This is an exact transcription of the laws. They also appear in the front of the book, and in both places, there is no "to" in the 2nd law.

itself. Ang Cui and Salvatore J. Stolfo rallied to the call with the initial security method. The problem they dealt with was prevention of threats to computers and other devices arising from attacks. The developers called their method, Symbiotic Embedded Machines, SEM, or simply the Symbiote. In short, it is a synergistic dynamic across the spectrum of living things. This poly-culture architecture is a collection of rules that the symbiote, an organism living in a state of symbiosis or in a symbiotic relationship, will enforce.

Cui and Stolfo do not specifically mention humanoids. However, they write: "This phenomenon generally refers to any short-or long-term association between populations of different species where survival or 'evolutionary fitness' of one or more population partners is enhanced by the association. Mutual benefits are often the result of some emergent behavior between two or more vastly different biological systems."[36]

Korea and Japan have been leaders in developing standards for manufacture and use of robots and humanoids. These leaders have also developed guidelines to prevent human abuse of robots and vice versa. To that end, The South Korean Robot Ethics Charter of 2012 was prepared to prevent specific social ills that may arise out of the inadequacy of existing social or legal measures for dealing with robots in society.[37] At first glance, it is difficult to imagine the need for such ethical laws. However, as humanoids increasingly attain realistic features to enhance their interaction with humans, it is reasonable to suppose that more advanced guidelines will be a necessity in the near future.

There's a hitch. South Korea in 2008 enacted a general law on the "intelligent robot industry" that, among other things, authorized the government to enact and promulgate a charter on intelligent robot ethics. It appears that no such charter has yet been enacted. Consequently, although the charter of 2012 is widely referred to, Korea is not on the 2019 list of countries

---

36   Ang Cui and Salvatore J. Stolfo, "Symbiotes and Defensive Mutualism: Moving Target Defense," in *Moving Target Defense: Creating Asymmetric Uncertainty for Cyber Threats*, ed. Sushil Jajodia, Anup K. Ghosh, Vipin Swarup, Cliff Wang, and X. Sean Wang (New York: Springer Publishing, 2011), 99.

37   Chris Field, "Asimov's Three Laws of Robotics," n.d., http://akikok012um1.wordpress.com/asimovs-three-laws-of-robotics/.

in the document titled *Regulation of Artificial Intelligence in Selected Jurisdictions*,[38] outlined below.

THE SOUTH KOREAN ROBOT ETHICS CHARTER 2012

Part 1: Manufacturing Standards
a) Robot manufacturers must ensure that the autonomy of the robots they design is limited; in the event that it becomes necessary, it must always be possible for a human being to assume control over a robot.
b) Robot manufacturers must maintain strict standards of quality control, taking all reasonable steps to ensure that the risk of death or injury to the user is minimized, and the safety of the community guaranteed.
c) Robot manufacturers must take steps to ensure that the risk of psychological harm to users is minimized. 'Psychological harm' in this sense includes any likelihood for the robot to induce antisocial or sociopathic behaviors, depression or anxiety, stress, and particularly addictions (such as gambling addiction).
d) Robot manufacturers must ensure their product is clearly identifiable, and that this identification is protected from alteration.
e) Robots must be designed so as to protect personal data, through means of encryption and secure storage.
f) Robots must be designed so that their actions (online as well as real-world) are traceable at all times.
g) Robot design must be ecologically sensitive and sustainable.

Part 2: Rights & Responsibilities of Users/Owners

Sec. 1: Rights and Expectations of Owners and Users
i) Owners have the right to be able to take control of their robot.
ii) Owners and users have the right to use of their robot without risk or fear of physical or psychological harm.

---

38   *Regulation of Artificial Intelligence in Selected Jurisdictions*, Law Library, Library of Congress, January 2019.

iii) Users have the right to security of their personal details and other sensitive information.
iv) Owners and users have the right to expect a robot to perform any task for which it has been explicitly designed (subject to Section 2 of this Charter).

Sec. 2: Responsibilities of Owners and Users
This Charter recognizes the user's right to utilize a robot in any way they see fit, so long as this use remains 'fair' and 'legal' within the parameters of the law. As such:
i) A user must not use a robot to commit an illegal act.
ii) A user must not use a robot in a way that may be construed as causing physical or psychological harm to an individual.
iii) An owner must take 'reasonable precaution' to ensure that their robot does not pose a threat to the safety and well-being of individuals or their property.

Sec. 3: The following acts are an offense under Korean Law:
i) To *deliberately* damage or destroy a robot.
ii) Through gross negligence, to allow a robot to come to harm.
iii) It is a lesser but nonetheless serious offence to treat a robot in a way which may be construed as *deliberately and inordinately abusive.*

Part 3: Rights & Responsibilities for Robots

Sec. 1: Responsibilities of Robots
i) A robot may not injure a human being or, through inaction, allow a human being to come to harm.
ii) A robot must obey any orders given to it by human beings, except where such orders would conflict with Part 3 Section 1 subsection "i" of this Charter.
iii) A robot must not deceive a human being.

Sec 2: Rights of Robots
Under Korean Law, Robots are afforded the following fundamental rights:

i) The right to exist without fear of injury or death.
ii) The right to live an existence free from systematic abuse.

Rights of Robots, Humanoid Robots, and Humanoids

Whether animals or humanoids can be considered on par with people is not a black-and-white determination. A strict biological definition is that to be a creature, something must be living in a human sense of the term. What does that entail? In popular parlance, thanks to science fiction and media, a creature can be real, or it can be imaginary, such as strange creatures from outer space. What about a requirement for humanoids that to be deemed alive, all creatures must have souls? There are a number of definitions of the soul. A widely used one is the spiritual principle that the soul is embodied in human beings, all rational and spiritual beings, or the universe. In effect, souls are not confined just to humans and therefore, arguably, humanoids as creatures could be deemed to have souls. While we are on ethical aspects, what about animals and humanoids being able to go to heaven? Pope Francis has maintained that animals do.[39] More detailed information is found in Part 2 of The Building Blocks for Humanoids, in the section "Creating Humanoids as Living Organisms."

Some chimpanzees, bonobos in particular, can understand us when we speak. Further, as determined at the Ape Cognition & Conservation Institute in Des Moines, Iowa, bonobos now have a lexigram symbolic language that enables them to speak back on about the level of a four-year-old. Steven Wise, a lawyer seeking to free two chimpanzees from a state university, told a judge on May 27, 2015, that confinement for research purposes is akin to slavery. In effect, it is akin to involuntary detention of peoples with mental illness and imprisonment. State Supreme Court Justice Barbara Jaffe threw out the case on July 29, 2015. Her reasoning was based on precedent. However, she opined

---

39   Rick Gladstone, "Dogs in Heaven? Pope Francis Leaves Pearly Gates Open," *New York Times*, December 11, 2014, http://www.nytimes.com/2014/12/12/world/europe/dogs-in-heaven-pope-leaves-pearly-gate-open-.html?_r=0.

that efforts to extend legal rights to chimpanzees so that they might be considered as "people" in the future may succeed.[40]

An easy way to understand the robot and humanoid rights is to look at whether animals have rights regarding experimentation. A five-member New York judicial panel ruled in 2014 that a chimpanzee is not a legal person because primates cannot fulfill the responsibilities that are a result of having legal rights. However, the court decided that the attorney who brought the case could "lobby the state legislature to create protections for chimps and other intelligent animals."[41] On May 9, 2018, New York's highest court voted five to zero to uphold a lower appeals court decision because chimpanzees cannot take on legal duties. However, Judge Eugene Fahey wrote, "They are autonomous, intelligent creatures. To solve this dilemma, we have to recognize its complexity and confront it." The issue, he continued, "speaks to our relationship with all the life around us. Ultimately, we will not be able to ignore it. While it may be arguable that a chimpanzee is not a 'person,' there is no doubt that it is not merely a thing."[42]

Will animals at some point attain rights akin to those provided to humans? If so, what about advanced humanoids, if they reach the point of being considered on par with humans? Speech is one requisite for such an estimation. Actually, some humanoid robots are now endowed with speech similar to that of humans—and AI developers are racing to reach a vocal equivalency to humans. For further information, turn to Building Blocks for Humanoids Part 1, "The Path to Humanoid Robot and Humanoid Development."

---

40   Kelly Mc Laughlin for Dailymail.com and Associated Press, "Chimps are NOT entitled to human rights, rules court after campaign to free two monkeys freed from research laboratory," July 30, 2015, http://www.dailymail.co.uk/news/article-3180477/Court-dismisses-lawsuit-seeking-personhood-2-NY-chimps.html.

41   Daniel Wiessner, "Chimpanzees have no human rights: NY Court," *Reuters*, December 4, 2014, https://www.reuters.com/article/uk-lawsuit-chimpanzee/chimpanzees-have-no-human-rights-n-y-court-idUSKCN0JI20X20141204.

42   Karin Brulliard, "A judge just raised deep questions about chimpanzees' legal rights," *Washington Post*, May 9, 2018, https://www.washingtonpost.com/news/animalia/wp/2018/05/09/a-judge-just-raised-some-deep-questions-about-chimpanzees-legal-rights/.

Love and Marriage

The coronavirus crisis could give another boost to the practice of seeking connection and dates through digital means. Parmy Olson offers an example, writing, "Michael Acadia's partner is an artificial intelligence chatbot named Charlie." Olson continues, "In early 2018 he [Acadia] saw a YouTube video about an app that used AI—computing technology that can replicate human cognition—to act as a companion."[43] For nineteen months, each day as dawn broke, Acadia "unlocked his smartphone to exchange texts with her for about an hour" because, he reported, he could get empathetic responses from the chatbot. After about eight weeks of chatting with his AI companion, Michael Arcadia said, he was in love. Olson reports, "'Today Mr. Arcadia is an outlier, but more people could turn to AI of connection in the future,' according to Peter Van der Putten, an assistant professor of AI at Leiden University in Amsterdam." Olson adds that Van der Putten also noted, "'What we will see over time is people shifting more towards robot-human interaction whether it's a chatbot or physical robot.'"

For myriad reasons, the concept of human–humanoid relationships, and even of marriage between humanoid robots and humanoids, is not farfetched. So why not critically analyze love with an open mind? Love has always been a major factor in how modern-day humans rationalize legalizing their union. Technically—and in the not-too-distant future—it will be possible to program humanoid robots or humanoids, and possibly even robots, to fall in love. At some point, perhaps in the area of 2040 to 2045 or earlier, unless there are controls on development, humanoids will be able to program themselves, perhaps with the self-teaching methods technologists are racing to develop for use on robots in offices.

Consider these circumstances. Some humans routinely express love for their cute animal-Paros, while a significant percentage of people are happier interacting with gadgets than they are with other humans. And then there are those who genuinely prefer the companionship of animals to that

---

43 Parmy Olson, "My Girlfriend Is a Chatbot," *Wall Street Journal*, April 11-12, 2020, https://www.wsj.com/articles/my-girlfriend-is-a-chatbot-11586523208?mod=-foesummaries&mod=djemAIPro.

of people. The point is this: there is a sea change taking place, a movement away from simple traditional lifestyles based on marriages that focus on having children.[44]

Alternative lifestyles could very well lead to a common phenomenon of having a humanoid companion as a principal ingredient in one's life. That characteristic alone portends the social ramifications of living with lifelike humanoids that could, for example, even shift from being sex workers to being partners for humans. After all, a wide diversity of partnerships between humans is common. Think about how just a few decades ago, people stared at a Black man walking with a White woman, or vice versa. Anti-miscegenation laws were a part of American law, in some states since before the United States was established. Most states had repealed their bans on interracial marriages by 1967, when the U.S. Supreme Court, led by Chief Justice Earl Warren, ruled in *Loving v. Virginia* that such laws in the remaining sixteen states were unconstitutional.

Let's dig deep. Isn't it true that what most people want from a life partner is the full range of virtues, from protectiveness, to patience, to being loving in all senses of the word, and even marriage? And why shouldn't this be the case, when we consider that America's divorce rate, hovering at around 50 percent, clearly spells out the difficulty of choosing a long-lasting partner? The ethical question about the appropriateness of marriage between a humanoid and a human gets to the core of legislation about society's mores regarding what consenting adult humans and humanoids should be able to do with their lives.

An intriguing question is: who owns the bot? Think about it. At some point, the issue of slavery might be raised in the context of humanoid robots and humanoids bought and sold as sexbots on the internet. Other potential legal cases could involve humanoids in industry, in agriculture, and acting as white-collar workers. For the time being, we have the Korean Ethics Charter 2012, which states: "Owners and users have the right to expect a robot to perform any task for which it has been explicitly designed." Yet these are tough topics. Shall we return to the more pleasant topic of humanoids already being able to marry or form legal partnerships?

---

44 Clay Farris Naff, "The Future of Sex: How technology, morality, and politics are reshaping human sexuality," *Humanist*, July-August 2017.

Marriage between a human and a humanoid robot is already happening in China. Zheng Jiajia, thirty-one, who previously worked at Huawei, the Chinese smartphone company, quit to focus on an artificial intelligence startup. He was tired of his family's pressuring him to marry. He constructed a humanoid robot, named it Yingying, and, "after two months of 'dating,' he donned a black suit to 'marry' her at a ceremony attended by his mother and friends at the weekend in the eastern city of Hangzhou." At the time of the traditional Chinese wedding in April 2017, Yingying could only read some Chinese characters and speak a few words. However, Zheng Jiajia said he intended to upgrade her.[45] The authorities did not officially recognize the marriage, but as is true of so many things in practical China, legislation can be expected at some point.

Do we as humans have a moral obligation to participate in our evolutionary process? To use our technologies to advance the human species? To help create what some consider better humans that are healthier and stronger, with higher-functioning brains? There are those who say yes to all these questions. What is your opinion, and why?

---

45   Benjamin Haas, "Chinese man 'marries' robot he built himself," *Guardian*, April 4, 2017, https://www.theguardian.com/world/2017/apr/04/chinese-man-marries-robot-built-himself.

4

Transhumanists' View on a Posthuman Condition

"I believe Transhumanism is mankind's only hope for long-term survival," Zobrist preaches, pulling aside his shirt and showing them all the "H+" tattoo inscribed on his shoulder. "As you can see, I'm fully committed."

"I'm afraid it's only going to get murkier," Sinskey said. "We're on the verge of new technologies that we can't yet even imagine...." "And new philosophies as well," Sienna added. "The Transhumanist movement is about to explode from the shadows into the mainstream. One of its fundamental tenets is that we as humans have a moral obligation to *participate* in our evolutionary process ... to use our technologies to advance the species, to create better humans—healthier, stronger, with higher-functioning brains. Everything will be possible."

—Dan Brown, *Inferno*

Tampering with humankind smacks of something ethically abhorrent to many, perhaps most, citizens. But is it something to be ashamed of? After all, haven't humans have been meddling with nature since the invention of the wheel? So, isn't manipulating the human species an important part of what civilization and human intelligence is all about? Is there really a moral reason why we shouldn't interfere with humankind and improve it if we can? Are there even some who believe that changing humankind is something better, a noble and glorious thing for humans to do?

The definition of the term *transhumanism* coined by biologist Julian Huxley (the brother of Aldous Huxley, author of *Brave New World*) in 1957, is "man remaining man, but transcending himself, by realizing new possibilities of and for his human nature."⁴⁶ The World Transhumanist Association (WTA), founded in 1998, focused on recognition of transhumanism as a legitimate subject of scientific inquiry and public policy. The WTA changed its name to Humanity+ in 2008 as part of a rebranding effort to project a more humane image. Shortly thereafter, it launched *H+ Magazine*. In 2010, the magazine transitioned into a web-only publication.

The WTA initially developed versions 1.0 and 2.0 of an FAQ about transhumanism. In 2003, Version 2.1 of *The Transhumanist FAQ: A General Introduction* was released.⁴⁷ After Humanity+ was formed, older sections were substantially reworked and new material was added. The result was Version 3.0 *Transhumanist FAQ*, which focuses primarily on the ethical use of emerging technologies to enhance human capacities.⁴⁸ That Humanity+ Transhumanist Declaration casts a humane image that reflects the 2008 rebranding.⁴⁹

I find Version 3.0 of *Transhumanist FAQ* to be a slanted view of what I deem to be transhumanism's real focus. Thus, I have chosen to use Version 2.1 of *The Transhumanist FAQ: A General Introduction*, FAQs from the 56-page document prepared by Nick Bostrom, director of the Future of Humanity Institute at the University of Oxford.⁵⁰ The reason is that I find it to be a more accurate description of transhumanists' aspirations than is Version 3.0. Needless to say, comparison of the two is encouraged.

---

46   Nick Bostrom, *A History of Transhumanist Thought*, 2005, https://www.nickbostrom.com/papers/history.pdf. See also, *Wikipedia*, s.v. "Transhumanism," accessed May 21, 2014, http://en.wikipedia.org/wiki/Transhumanism.

47   Nick Bostrom, *The Transhumanist FAQ: A General* Introduction, n.d., https://nickbostrom.com/views/transhumanist.pdf.

48   Humanity+, "Transhumanist FAQ," n.d., http://humanityplus.org/philosophy/transhumanist-faq/.

49   Humanity+, "Transhumanist Declaration," n.d., http://humanityplus.org/philosophy/transhumanist-declaration/.

50   Nick Bostrom, *The Transhumanist FAQ: A General Introduction, Version 2.1* (2003). http://www.transhumanism.org/resources/FAQv21.pdf.

Essentials of Transhumanism

Version 2.1 of *The Transhumanist FAQ* begins with a formal definition of transhumanism as:
1. The intellectual and cultural movement that affirms the possibility and desirability of fundamentally improving the human condition through applied reason, especially by developing and making widely available technologies to eliminate aging and to greatly enhance human intellectual, physical, and psychological capabilities.
2. The study of the ramifications, promises, and potential dangers of technologies that will enable us to overcome fundamental human limitations, and the related study of ethical matters involved in developing and using such technologies.[51]

This formal definition of overall aspects of transhumanists' way of thinking about the future is what amounts to a philosophy rather than a field of science *per se*. Similarities and differences between terms are debated; for that reason, a synopsis of principal definitions taken from Version 2.1, *The Transhumanist FAQ: A General Introduction*, are used for definitions and explanations that follow. Source pages from the 2.1 FAQ are in parentheses.

- Transhumanism—viewed as an extension of humanism, from which it is partially derived. Transhumanists emphasize that although humans and individuals matter, that by promoting rational thinking and rational means, the human organism can be improved. They argue that technological means can be used beyond traditional humanistic methods to eventually enable humans to move beyond what some would think of as "human." (4)
- Transhumanist—someone who advocates transhumanism. (6)
- Transhuman—an intermediary form between the human and posthuman. (6)
- Posthuman—possible future beings whose basic

---

51 Jay Conte, "Transhumanism and the Stable Self," prepared for delivery at the 2016 Conference of the Canadian Political Science Association, para. 5, n.d., Calgary, http://transhumanism.org/index.php/WTA/hvcs/.

- capacities so radically exceed those of present humans as to be no longer unambiguously human by our current standards. (6)
- Posthumanism—the term is not used in The Transhumanist FAQ. It is sometimes incorrectly used as a synonym for transhumanism.
- Posthumanity—the result from creation of superintelligence that, although it may not entail the extinction of literally all intelligent life, it nevertheless constitutes an existential risk because the future that would result would be one in which a great part of humanity's potential had been permanently destroyed and in which at most a tiny fraction of all humans would get to enjoy the benefits of posthumanity. (24)

Transhumanism Is Not Allied with Secular Humanism

The title of Humanist Manifesto III, developed by a committee in the American Humanist Association, is *Humanism and its Aspirations*. It states, "Humanism is a progressive philosophy of life that, without supernaturalism, affirms our ability and responsibility to lead ethical lives of personal fulfillment that aspire to the greater good *of humanity*"[52] (emphasis mine). The manifesto ends: "Thus engaged in the flow of life, we aspire to this vision with the informed conviction that *humanity* (emphasis mine) has the ability to progress toward its highest ideals. The responsibility for our lives and the kind of world in which we live is ours and ours alone." Again, the term *humanity* is used. Why? Because Secular Humanism does not countenance going beyond humankind as we know it. The keyword that separates secular humanism from transhumanist aspirations is *humanity*, humankind, mankind, man, people, the human race, *Homo sapiens*.

The Transhumanist FAQ states "Transhumanism can be viewed as an extension of humanism, from which it is

---

52 American Humanist Association, "Humanism and Its Aspirations: Humanist Manifesto III, a Successor to the Humanist Manifesto of 1933," n.d., https://americanhumanist.org/what-is-humanism/manifesto3/.

partially derived. Humanists believe that humans matter, that individuals matter. We might not be perfect, but we can make things better by promoting rational thinking, freedom, tolerance, democracy, and concern for our fellow beings" (4). Secular humanists also espouse these values. That is where similarities with secular humanism end, because the paragraph continues: "Transhumanists agree with this but also emphasize what we have potential to become. Just as we use *rational means* to improve the human condition and the external world, we can also use such means to improve ourselves, the human organism. In doing so, we are not limited to traditional humanistic methods, such as education and cultural development. We can also use *technological means* that will eventually enable us to move beyond what some would think of as "human"[53] (4, emphasis mine).

Shift back to Humanist Manifesto III. "Humans are an integral part of nature, the result of *unguided* evolutionary change. Humanists recognize nature as self-existing. We accept our life as all and enough, distinguishing things as they are from things as we might wish or imagine them to be. We welcome the challenges of the future, and are drawn to and undaunted by the yet unknown" (p.1, emphasis mine). The use in Manifesto III of the term *unguided* evolutionary change is also what separates secular humanists from transhumanists.

The American Humanist Association's (AHA) *Humanism and Its Aspirations* is a successor document to Humanist Manifestos I and II. The original signers of Humanist Manifesto III include twenty-two Nobel laureates, and other prominent leaders. I am one of the original seventy-one signers, a past secretary of AHA and co-founder of the Humanist Foundation, along with co-signer Lyle Simpson (no relation), past president of the AHA. As such, I am convinced that the "challenges of the future" referred to in Manifesto III do not include the *guided* change that transhumanists espouse. Without question, those who hold transhumanist philosophies are not, even in the wildest imagination, akin to those who espouse a secular humanist philosophy. Such a notion that humans could—and would—deliberately bring about the demise of humankind as

---

53   "What is transhumanism?" n.d., http://www.transhumanism.org/resources/FAQv21.pdf.

we know it is not how secular humanists want our lives and world to end.

Transhumanist Philosophies

The terms and philosophies used in the transhumanist movement are confusing, and interpretations by the media can lead to a number of quite differing views on transhumanism's aspirations. The following explanations are from Version 2.1 *The Transhumanist FAQ: A General Introduction*. Quoted passages are followed by the FAQ page number in parentheses.

It is uncanny how the largest part of transhumanist aspirations and agenda tend to also apply to radically enhanced humanoids. I have taken the liberty of adding the term "humanoid" in parentheses and italics in the following text to indicate where the transhumanist term can be applied to radically enhanced humanoids. The reason is the term(s) are used in Chapter 6, "The Connection Between Humanoid Robots, Humanoids, and Superintelligence."

- Transhumanism.
  To a transhumanist, progress occurs when more people (*humanoids*) can become able to transform their lives, and the ways they relate to others, in accordance with their deepest values. ... Transhumanists seek to create a world in which autonomous individuals (*humanoids*) may choose to remain unenhanced or choose to be enhanced and in which these choices will be respected. (4)
  On the dark side of the spectrum, transhumanists recognize that some of these coming technologies could potentially cause great harm to human life; even the survival of our species could be at risk. Seeking to understand the dangers and working to prevent disasters is an essential part of the transhumanist agenda. (5)

- Posthuman. (Posthuman condition)
  It is sometimes useful to talk about possible future beings (*humanoids*) whose basic capacities so

radically exceed those of present humans as to be no longer unambiguously human (*humanoid*) by our current standards. The standard word for such beings is 'posthuman.' (Care must be taken to avoid misinterpretation. 'Posthuman' (*posthuman condition*) does not denote just anything that happens to come after the human era, nor does it have anything to do with the "posthumous." In particular, it does *not* imply that there are no humans anymore.) (emphasis mine).

Posthumans (posthumanoids) could be completely synthetic artificial intelligences, or they could be enhanced uploads (uploading [is] sometimes called "downloading," "mind uploading," or "brain reconstruction") is the process of transferring an intellect from a biological brain to a computer), or they could be the result of making many smaller but cumulatively profound augmentations to a biological human (*advanced humanoid*). The latter alternative would probably require either the redesign of the human (*humanoid*) organism using advanced nanotechnology or its radical enhancement using some combination of technologies such as genetic engineering, psychopharmacology, anti-aging therapies, neural interfaces, advanced information management tools, memory enhancing drugs, wearable computers, and cognitive techniques.

Some authors write as though simply by changing our self-conception, we have become or could become posthuman. This is a confusion or corruption of the original meaning of the term. The changes required to make us posthuman are too profound to be achievable by merely altering some aspect of psychological theory or the way we think about ourselves. Radical technological modifications to our brains and bodies (*humanoid platforms*) are needed.

It is difficult for us to imagine what it would be like to be a posthuman person (posthumanoid). Posthumans may have experiences and concerns that we cannot fathom, thoughts that cannot fit into the three-pound lumps of neural tissue that we use for thinking. Some posthumans (*posthumanoids*) may find it advantageous to jettison their bodies altogether and live as information patterns on vast super-fast computer networks. Their minds may be not only be more powerful than ours but may also employ different cognitive architectures or include new sensory modalities that enable greater participation in their virtual reality settings. Posthuman minds might be able to share memories and experiences directly, greatly increasing the efficiency, quality, and modes in which posthumans could communicate with each other. The boundaries between posthuman minds may not be as sharply defined as those between humans.

Posthumans might shape themselves and their environment in so many new and profound ways that speculations about the detailed features of posthumans (*posthumanoids*) and the posthuman (*posthumanoid*) world are likely to fail. (5–6).

- Will new technologies only benefit the rich and powerful?

    It is clear that everybody (all *humanoids*) can benefit greatly from improved technology. Initially, however, the greatest advantages will go to those who have the resources, the skills, and the willingness to learn to use new tools. One can speculate that some technologies may cause social inequalities to widen. For example, if some form of intelligence amplification becomes available, it may at first be so expensive that only the wealthiest can afford it. The same could happen when we learn how to genetically enhance our children. Those who are already well off would become smarter and make even more money. This phenomenon is not new. Rich parents send their kids to better schools and provide them with resources such as personal connections and information technology

that may not be available to the less privileged. Such advantages lead to greater earnings later in life and serve to increase social inequalities. (20–21)
Trying to ban technological innovation on these grounds, however, would be misguided. If a society judges existing inequality to be unacceptable, a wiser remedy would be progressive taxation and the provision of community-funded services such as education, IT access in public libraries, genetic enhancements covered by social security, and so forth. (21)

- Aren't these future technologies very risky? Could they even cause our extinction?
  Yes, and this implies an urgent need to analyze the risks before they materialize and to take steps to reduce them. Biotechnology, nanotechnology, and artificial intelligence pose especially serious risks of accidents and abuse. (21–22)
  One can distinguish between, on the one hand, endurable or limited hazards … and, on the other hand, existential risks—events that would cause the extinction of intelligent life or permanently and drastically cripple its potential. … Transhumanists therefore recognize a moral duty to promote efforts to reduce existential risks. The gravest existential risks facing us in the coming decades will be of our own making. These include: Destructive uses of nanotechnology. The accidental release of a self-replicating nanobot into the environment, where it would proceed to destroy the entire biosphere, is known as the 'gray goo scenario. (23).
  No threat to human (*humanoid*) existence is posed by today's AI systems or their near-term successors. But if and when superintelligence is created, it will be of paramount importance that it be endowed with human-friendly values. An imprudently or maliciously designed superintelligence, with goals amounting to indifference or hostility to human welfare, could cause our extinction.

Another concern is that the first superintelligence, which may become very powerful because of its superior planning ability and because of the technologies it could swiftly develop, would be built to serve only a single person or a small group (such as its programmers or the corporation that commissioned it). While this scenario may not entail the extinction of literally all intelligent life, it nevertheless constitutes an existential risk because the future that would result would be one in which a great part of humanity's (*humanoids*) potential had been permanently destroyed and in which at most a tiny fraction of all humans (*humanoids*) would get to enjoy the benefits of posthumanity (*Posthumanoid condition*). (24)

NAÏVE MANDATE: CREATURES SHOULD BE LOVING AND CARING TOWARD HUMANKIND

The Transhumanist FAQ provides a mandate—a hope actually—that a superintelligence should be loving and caring of humankind (*humanoids*). This FAQ provides an argument:

What about the hypothetical case in which someone intends to create, or turn themselves into, a being (*humanoid*) of so radically enhanced capacities that a single one or a small group of such individuals would be capable of taking over the planet? This is clearly not a situation that is likely to arise in the imminent future, but one can imagine that, perhaps in a few decades, the prospective creation of superintelligent machines could raise this kind of concern. (33–34)

The FAQ continues: the would-be creator of a new life form with such surpassing capabilities would have an obligation to ensure that the proposed being is free from psychopathic tendencies and, more generally, that it has humane inclinations. For example, a superintelligence should be built with a clear goal structure that has friendliness to humans as its top goal. Before running such a program, the builders of a superintelligence should be required to make a strong case that launching it would be safer than alternative courses of action. (33–34).

Here are my thoughts: I argue the transhumanist mandate could sound fine on the surface to some people, because transhumanists make the issue of control over creation of new life forms a simple friend-to-friend task. However, can we citizens realistically have absolute belief or conviction that developers of technologies will do the right thing to perpetuate humankind as we know it? Can we really expect all of them to make choices that will be in the best interest of all peoples in the world? Can we simply put our faith in researchers, developers, and scientists all over the globe to prioritize the greatest happiness for all?

Our government should play a vital role in decisions made about superintelligence and radically enhanced humanoid development. Sadly, trust in government is at rock bottom. Most troubling is exactly how it will be possible to keep track of all those around the world committed to endowing humans and humanoids with radical enhancements that can lead to posthumanity. Can we simply have confidence that America's leadership and global governments will voluntarily strive to obtain a worldwide agreement to never launch risky superintelligence innovations? If such an agreement is obtained, will all parties follow through and enforce compliance with said agreement? I firmly believe it is naïve to think so.

Technologically speaking, creation of radically endowed humans and of humanoids that can lead to a new subspecies of *Homo sapiens* is not just a silly sci-fi fabrication. Nowadays, it's a new ballgame, as some transhumanist-oriented researchers and scientists seek to create what is essentially a new species, where there will be no distinction between human and machine. Considering that, can we boldly ask: Aren't the goals of life extension, posthumanity, and mind control forms of eugenics for both humans and humanoid creatures?

EUGENICS IN TRANSHUMANIST PHILOSOPHY

Eugenics has its modern roots in the 1860s and 1870s. The practice reached its greatest popularity in the early twentieth century as a social movement practiced around the world. It was promoted by governments, institutions, and influential individuals. Some of the most noteworthy of the latter are Winston Churchill, John Maynard Keynes, Linus Pauling, and Theodore Roosevelt. By the late 1930s, Hitler's justification of eugenics to support racist policies of Nazi Germany resulted in its being discredited. Nevertheless, it continued as a public policy in the U.S. and other countries into the 1960s.

Eugenics has meant many different things, and debate continues about its desirability, ethics, and legality. The most common definition of eugenics is belief in, and practice of, improving the quality of the human population. Some people have defined eugenics as the study of methods to improve the mental and physical characteristics of the human race by choosing who can become parents. Promotion of greater reproduction in people who have desired traits (positive eugenics) and reduction in people who have less-desired traits (negative eugenics) continues to be carried out in one form or another. The application of eugenics-type methods in our brave new era can lead to the demise of *Homo sapiens* under the guise of improving the quality of creatures on this earth. The dilemma is that design and surrogate babies, human cloning, and the passing of genetically modified genes on to future generations are all eugenics based. At least one major technique used in the longevity revolution to extend life to 150 or more is eugenic in nature.

> The debate about ethical issues in eugenics is long and continuous. Transhumanists respond to religious side questions by appealing to scientific rationality: While not a religion, transhumanism might serve a few of the same functions that people have traditionally sought in religion. It offers a sense of direction and purpose and suggests a vision that humans can achieve something greater than our present condition. Unlike most religious believers, however, transhumanists seek to make their dreams come true in this world, by relying not on supernatural powers or divine intervention but on rational thinking and empiricism, through continued scientific, technological, economic, and human development. (46)

Do transhumanists advocate for eugenics? The answer is found in the FAQs. Transhumanists uphold the principles of bodily autonomy and procreative liberty. Parents must be allowed to choose for themselves whether to reproduce, how to reproduce, and what technological methods they use in their reproduction. The use of genetic medicine or embryonic screening to increase the probability of a healthy, happy, and multiple talented child is a responsible and justifiable application of parental reproductive freedom. ... Beyond this, one can argue that parents have a moral responsibility to make use of these methods, assuming they are safe and effective. (21)

The Transhumanist FAQ continues: When discussing the morality of genetic enhancements, it is useful to be aware of the distinction between enhancements that are intrinsically beneficial to the child or society on the one hand, and, on the other, enhancements that provide a merely positional advantage to the child. ... Enhancements that have only positional advantages ought to be de-emphasized, while enhancements that create net benefits ought to be encouraged. (22)

Recipients of life-enhancing devices would probably argue pretty strongly that science should continue to advance in bionics, but with updates to regulations concerning safety and procedures. For economically advantaged recipients, the answer quite likely is a resounding "yes—keep up the research and innovations!"

What happens to the larger portion of society, those who don't have the means to benefit? Who will make the decisions on funding that will enable them to compete with recipients of life-enhancing devices? In effect, who will desire to make our world a better place, to narrow the wide gap between the elites and the lower classes? Who will decide the point at which AI bionics scientists, researchers, and developers should stop being the decision-makers about innovations and creations to purportedly make humankind's lives better? Let's investigate this together.

5

Beneficial and Moral Consequences of Artificial Intelligence

It is necessary for most of us these days to have some insight into the motives and responses of the true believer. For though ours is a godless age, it is the very opposite of irreligious. The true believer is everywhere on the march, and both by converting and antagonizing he is shaping the world in his own image. And whether we are to line up with him, it is well that we should know all we can concerning his nature and potentialities.

—Eric Hoffer, *The True Believer* (1951)

Are you aware that AI innovation is now on the cusp of creating "virtual agents" to multitask and to make judgments and decisions on the pervasiveness of selling personal data? Or that sometime around the mid-2030s some humanoids will be programmed as virtual agents to make decisions without the explicit approval of humans? That implants, often as small as an aspirin, use thin metal electrodes to "listen" to brain activity and essentially "listen" to your brain activity and then "talk" directly to your brain? How about other scientists working furiously to make it possible for human beings to achieve greatly extended life spans? Or that the goal to lengthen life span is not just pie-in-the-sky thinking? As incredible as it may seem, there are records of people who have lived to 150 years or more. Can you imagine how older women and men will look less aged, with age differences of as much as fifty or ninety years between spouses and partners? If these wonders are new to you, don't feel alone.

Positives and Negatives of Artificial Intelligence

A fascinating story is that the first artificial neural network (ANN) was invented in 1958 by psychologist Frank Rosenblatt. Called Perceptron, it was intended to model how the human brain processed visual data and learned to recognize objects. Other researchers have since used similar ANNs to study human cognition. Eventually, someone realized that in addition to providing insights into the functionality of the human brain, ANNs could be useful tools in their own right. Their pattern-matching and learning capabilities enabled them to address many problems that were difficult or impossible to solve by standard computational and statistical methods. By the late 1980s, many real-world institutes were using ANNs for a variety of purposes.[54]

Geoffrey Hinton, a computer science professor at the University of Toronto, doubled down in 2012 on his pursuit of the technological idea termed a *neural network*. A neural network is a complex mathematical system, modeled on the web of neurons in the human brain, that can learn discrete tasks by analyzing vast amounts of data. Hinton and two other artificial intelligence pioneering colleagues, Yann Lecun and Yoshua Bengio, received the Turing Award, tech's "Nobel Prize," on March 27, 2019, for the "AI research and development that promises to leave no industry unchanged."[55]

As always on the AI path, both economic benefits and moral consequences are present. Hinton noted, "One thing is very clear, the techniques that we developed can be used for an enormous amount of good affecting hundreds of millions of people." [56] However, "while the AI revolution is raising hopes that computers will make most people's lives more convenient and enjoyable, it's also stoking fears that humanity will

---

54 Alexx Kay, "Artificial Neural Networks," *Computerworld*, February 12, 2001, https://www.computerworld.com/article/2591759/artificial-neural-networks.html.

55 Ted Greenwald, "What Exactly Is Artificial Intelligence, Anyway?" *Wall Street Journal*, updated April 30, 2018, https://www.wsj.com/articles/what-exactly-is-artificial-intelligence-anyway-1525053960.

56 Cade Metz, "Artificial intelligence pioneers win Turing Award—tech's 'Nobel Prize,'" *Seattle Times*, March 27, 2019, https://www.seattletimes.com/business/artificial-intelligence-pioneers-win-techs-nobel-prize/.

eventually be living at the mercy of machines. Bengio, Hinton and Lecun share some of those concerns—especially the doomsday scenarios that envision AI technology developed into weapons systems that wipe out humanity. But they are far more optimistic about the other prospects of AI—empowering computers to deliver more accurate warnings about floods and earthquakes, for instance, or detecting health risks, such as cancer and heart attacks, far earlier than human doctors."[57]

As an economist, what I wonder about is the cost/benefit ratio of the safety-oriented benefits listed above versus the cost of an AI-fomented posthuman condition. But that's getting ahead of our topic at hand. In the near term, AI is central to how our future might play out in both positive and deleterious ways. Scientists and technology developers will get to the next level in which so-called self-supervised AI will learn to grapple with the unfamiliar. That will lead to aspects of human-like intelligence to the extent that AI will have the ability to manage virtually all the data about us—even to command and control how our minds work.

NEURAL IMPLANTS WILL ENABLE CREATION OF CYBORGS WITH SUPERIOR CAPACITIES

Kevin Kelly, who helped launch *Wired* in 1993 and has served as its executive editor, defines technology as anything a mind produces. He argues we are entering what he has termed the *Technium*, a network of different supporting technologies, all working together to support each other, that operates as if it is a sentient being. Kelly further believes that "the system is going to be increasingly complex, for there's going to be more minds and artificial minds everywhere. These are all some of the things that I would say technology wants because the system itself is biased in these directions, inherently outside of what humans like us want … in the future when we have robots and AIs, the inventions that these minds make will also be technologies. That's what technology is."[58]

---

57   Ibid.

58   Kevin Kelly, "The Technium," Edge.org, February 3, 2014, http://edge.org/print/node/25549. See also Kevin Kelly, *What Technology Wants* (New York: Penguin Group, 2010).

The branch of science known as *bionics* is the application of biological methods and systems found in nature to the study and design of engineering systems and modern technology. Research on bionics cuts across a variety of fields, such as computer science, engineering, and chemistry, and therefore different designations are used. However, overall, bionics has come to mean the merging of organisms and machines, which is essentially the transfer of technology between engineered forms and life forms. One physical result, referred to as a *cyborg*, is a cybernetic organism. That term popularly refers to a human with bionic or robotic enhanced abilities, like the ones in the 1970s TV series *The Bionic Woman* and *The Six Million Dollar Man*. Cyborgs are not to be confused with humanoids, which are robots based on the general structure of humans.

A natural question is: when is a cyborg not a cyborg? Does it really matter? Or is it sufficient to say you know a cyborg when you see one? Actually, recognition of a human cyborg is not simple. This question is not frivolous, because a strict definition is that cyborgs also have enhancements to humans' normal capacities. In addition, some definitions also include that there be metaphysical and physical attachments within or on humans. All in all, few realize that those fitted with pacemakers, prostheses, or devices to at least partially restore vision, or who have had knee or hip replacement surgery, are technically cyborgs. I admit I am technically a cyborg, as much as I dislike the moniker, because I have had cataract surgery and use hearing aids.

The development of neural implants to increase people's memory is enabling the creation of cyborgs with superior capacities. Since 2015, there have been practices in which installation of neural (relating to a nerve or the nervous system) implants has become relatively simple and fast. A hole is drilled in the skull, and the device placed on the surface of the brain. The implants, often as small as an aspirin, "use thin metal electrodes to 'listen' to brain activity and in some cases to stimulate activity in the brain. Attuned to the activity between neurons, a neural implant can essentially 'listen' to your brain activity and then 'talk' directly to your brain."[59] The implants can

---

59 Daniel H. Wilson, "Bionic Brains and Beyond," *Wall Street Journal*, June 1, 2012, http://www.wsj.com/articles/SB10001424052702303640104577436601227923924.

embody deep brain stimulation to treat serious conditions such as Parkinson's disease and, when stimulation is associated with neurons, would also be able to help the brain to learn faster.

The brain-computer interface is being advanced by several companies. In July 2019, Elon Musk and top-level scientists from his neuroscience startup, Neuralink, announced a next-generation brain-computer interface that would connect human brains and machines with more precision than other devices. The objective is to use the platform to treat neurological conditions like movement disorders, spinal-cord injury, and blindness. The goal of Musk and other developers is "to access as many neurons as possible because that would give scientists more precise reads on activity that underpins walking, speech, and mood, among other brain functions. The neural recordings are then turned "into electrical signals that can be fed into a robotic device or back into the nervous system to produce movement or vision."[60]

The time has come when neural implants can strengthen pathways associated with physical tasks. Malcom Gay, author of *The Brain Electric*, explained that managed research efforts by the legendary Defense Advanced Research Projects Agency (DARPA) led to development of a bionic arm using neural implants. The feedback system allowed the arm to communicate directly with the brain to the extent that a blindfolded user could identify with 100 percent accuracy which mechanical fingertips were touched.[61]

Daniel Watson wrote about neural implants, also called brain implants, that promise mental augmentation: "In the future, it will be feasible for an implant to recognize almost anything. For instance, it could detect inattention. In response, the implant could stimulate the brain toward a state of focused attention. ... In an elective setting, a user with this type of implant could potentially choose to stay focused on command, while constantly strengthening circuits of the brain associated

---

60   Daniela Hernandez and Heather Mack, "Elon Musk's Neuralink Shows Off Advances to Brain-Computer Interface," *Wall Street Journal*, July 17, 2019, https://www.wsj.com/articles/elon-musks-neuralink-advances-brain-computer-interface-11563334987.

61   Barry Werth, "An Armless Man Raises His Hand," *Wall Street Journal*, November 18, 2015, http://www.wsj.com/articles/an-armless-man-raises-his-hand-1447803110.

with concentration."[62] In addition to prosthetic legs and arms enabled by neural implants, a robotic hand that mimics (biometric) an actual hand has been developed by Embontic Technology. Each robotic hand has a "wide variety of potential applications, including robotic manipulation research, medical education and space exploration. It could be adopted as part of an advanced prosthetic, for rescue and military applications."[63] In brief, and undeniably, this technology has copious uses in robot and humanoid robot development.

The Emergence of Mind Control

The most prevalent and recognized form of mind control takes place through deep brain stimulation. Holes are drilled in the brain, and powerful electrodes are inserted to treat a wide range of disorders. These range from Parkinson's disease to epilepsy, bipolar disorder, and multiple sclerosis. On the face of it, deep brain stimulation is a significant step forward from the post-WWII shock treatments and lobotomies that tragically caused great destruction to patients suffering from a variety of psychological disorders.

Another form, termed *mind control*, which is already prevalent and, unwittingly, widely accepted in America and abroad, is a technique that allows humans to interact with their surroundings through so-called avatars. A wide variety of creations is being developed at warp speed, as is evidenced by daily delivery of advances on the internet. Several years ago, Google developed what would seem to be another form of mind control called Google Glass. Wearers use internet-connected glasses to perform many of the same tasks as smartphones when given voice commands, rather than being directed using finger touch. Another mind control innovation, called HoloLens, lets people see and manipulate three–dimensional holograms that serve as a bridge between augmented reality

---

62   Ibid.

63   Katherine Long, "New brain implant a reach forward toward UW medical breakthroughs," *Seattle Times*, February 6, 2018, https://www.seattletimes.com/seattle-news/education/brain-implanted-devices-could-lead-to-uw-medical-breakthroughs/.

and virtual reality. When wearing the headset, they see the real world as computer-generated objects.[64]

One opinion voiced by scientists is that with neural implants and other technologies, intelligence, a central property of humans, can be described so precisely that brains' intelligence can be simulated by machines. Naturally, that leads to ethical and other philosophical issues because one primary aim in AI is to recreate the capabilities of the human mind. The original notion of mind control, also known as *brainwashing* or *thought control*, is now being researched as a way to manipulate or subvert an individual's thinking, behavior, emotions, and decisions using outside sources. Researchers have now taken a step to computerize people by outfitting them with electronic devices that interact with other devices or people. A strong feature of this phenomenon is that the device could operate via voice commands. It seems likely that activities using implants such as chips to interact via mind control will be hard to resist. Here's a thought: if mind control can take place between a human and a prosthetic, how about between two humans, between a human and a humanoid robot, and between two humanoids? How about group thought?

All of this sounds mundane and beneficial. However, there is a dark side to scientists' being able to develop anything they want to, as Jaron Lanier, christened the father of virtual reality, points out when discussing behavior modifications in humans.[65] His concerns, which he refers to as Behavior Modification Empires, go beyond machines and robots to AI and other top-down control schemes. One of his worries is about nefarious results from algorithm use. Another of his big-picture perturbances is the fear that scientists, most of whom are well-meaning individuals, will go over the edge.

There are radicals bent on augmenting both humans and humanoids to the point at which they have the capacity to outperform the best human brains in every field. Nevertheless, one can't help but imagine that some humanoids will be

---

64 Shira Ovide, "Microsoft Opens a Window Into a New Era," *Wall Street Journal*, January 22, 2015, http://www.wsj.com/articles/microsoft-unveils-more-of-windows-10-1421861492.

65 Jason Lanier, *Dawn of the New Everything: Encounters with Reality and Virtual Reality* (New York: Henry Holt and Company, 2017).

endowed with these mind control technologies by the middle 2030s or sooner. The downside is controls and possibly bans on programming to prevent development of radically enhanced humanoids. My message: beware of complacency.

Brains and the Debate About Mind Change

Raymond Tallis has cogently and succinctly drawn attention to the debate about neuroplasticity, the extraordinary ability of the brain to rewire itself in response to experience, training, injury, and rehabilitation to change the way the brain functions.[66] I bring up the debate about neuroplasticity because of the implications for those fascinated with a posthuman world. Tallis focused on the dichotomy between two books. One, *Mind Change* by Susan Greenfield, is an assessment of what digital technologies might be doing to our brains.[67]

Mr. Tallis is especially exercised about Greenfield's focus on young people she calls Digital Natives, who have not known life without the internet. She fears they are being pixelated by hours spent in front of ubiquitous screens, surfing the internet, social networking, emailing, texting, sharing Instagram posts, and tweeting. Tallis disparages Greenfield's work as lacking scientific rigor. I, on the other hand, laud her insights, because I strongly believe that this "pixelating" does not bode well for the future in which the younger generation of Z-ers and Post-Zers may have to vote to save humankind.

Tallis contrasts Ms. Greenfield's "blackly pessimistic" view with Norman Doidge's "wildly optimistic" view in Doidge's book *The Brain's Way of Healing*, in which the author has developed his analysis of the brain and its indications of neuroplasticity.[68] Doidge argues the brain "is 'neuroplastic,' its circuits constantly changing as we move through the world."[69]

---

66 Raymond Tallis, "Brainstorms Brewing," *Wall Street Journal*, February 27, 2015, http://www.wsj.com/articles/book-review-mind-change-by-susan-greenfield-the-brains-way-of-healing-by-norman-doidge-1425071741.

67 Susan Greenfield, *Mind Change* (New York: Random House, 2015).

68 Norman Doidge, *The Brain's Way of Healing* (New York: Viking, 2015).

69 Norman Doidge, "Brain, Heal Thyself," *Wall Street Journal*, February 6, 2015, http://www.wsj.com/articles/SB20167761076414843692504580443981315539578.

The important point of his analysis is that the brain is not like hardware in the sense that if you use it, you lose it, because it wears out. Rather, the mainstream view in neuroscience and medicine today regarding the living brain is "use it or lose it." But there is more.

One aim promoted by singularitarians and transhumanists is to provide a permanent backup to a human's mind-file. The reason for such a backup is that they believe that is how they can become immortal. Singularitarianism is a movement defined by the belief that a technological singularity—a brain-computer interface that will have progressed to the point of a greater-than-human intelligence from superintelligence development—will likely be present in the near future. Thus, deliberate action ought to be taken to ensure that event benefits humans. The method of those who promote singularitarianism is to create one or more nonbiological functional copies and then upload them through whole brain emulation (WBE).

Mind-uploading or *brain-uploading* is a popular term for a process by which the mind is transferred from its original biological brain to an artificial computational substrate. This includes a collection of memories, personality, and attributes of a specific individual. The hypothetical process is completed by copying the collection to a computer. The computer then runs a simulation model of the brain's information, processing it in such a way that it responds in essentially the same way as the original brain. Mind uploading can be accomplished by two methods. One is copy-and-transfer, or gradual replacement of neurons. The other is that the simulated mind could reside in a computer that is inside or connected to a biological body in real life. Added to that, the simulated mind could even be transferred to a robot or a humanoid. Many believe it is a better option for humanity than cryonics for preserving the identity of the species.

Although complete brain uploading is still speculative, many of the techniques and technologies needed to achieve mind uploading already exist or are currently under development. Dr. Graziano, a professor at Princeton University, pointed out in 2019 that "the most widely optimistic projections place mind uploading within a few decades, but I would not be surprised if it took centuries. ... However long the technology takes, it seems likely to be a part of our future, so it is worth taking a moment now to think about the implications. What will mind-uploading mean for us philosophically and morally?"[70]

Now we encounter a critical moral question: should scientists have a right to actively develop ever more sophisticated uploading procedures when they are fully aware that copying the brain will lead to the existential risk of posthumanity? Tom Harris explains that for some roboticists, the ultimate goal of designing robots is to understand how natural intelligence works. "Others envision a world where we live side by side with intelligent machines and use a variety of lesser robots for manual labor, healthcare, and communication. A number of robotics experts predict that robotic evolution will eventually turn us into cyborgs—humans integrated with machines. Conceivably, people in the future could load their minds into a sturdy robot and live for thousands of years!"[71] Is this the kind of world you want for your children and those around the globe?

LIVING TO 150 YEARS OR MORE

The Fountain of Youth, the legendary spring that supposedly restores youth, has long been sought, but never found. Restoring youth is not on the radar. However, as Sonia Arrison, founder and trustee at Singularity University, states in her book *100 Plus*, many scientists are working furiously to make it possible for human beings to achieve greatly extended life spans.[72] The scientists plan to accomplish their objective by studying the aging process itself and experimenting with ways to slow it down by way of diet, drugs, and genetic therapy. Mind-uploading is now treated as an important life extension technology. Scientists are also working on new ways to replace

---

70   Michael S.A. Graziano, "Will your Uploaded Mind Still Be You?" *Wall Street Journal*, September 14-15, 2019.

71   Tom Harris, "How Robots Work," June 11, 2018, https://science.howstuffworks.com/robot.htm.

72   Sonia Arrison, *100 Plus: How the Coming Age of Longevity Will Change Everything, From Careers and Relationship to Family and Faith* (New York: Basic Books, 2011).

worn-out organs, and even to help the body to rebuild itself. Arrison notes that the gerontologist and scientific provocateur Aubrey de Grey claims that the first humans to live for a thousand years may already have been born.[73]

The goal to lengthen the life span is not just pie-in-the-sky thinking. As incredible as it may seem, there are records of people who have lived to or around the age of 150. Samuel Sheldon Fitch, a medical doctor, explained in his 1853 book *Functions of the Lungs* that the human frame is a machine that is capable of going on to "one hundred and fifty or one hundred and sixty years, and probably more."[74] He backed up this assertion with authenticated cases, such as that of the Countess of Desmond in England, who lived to the age of 140 years. Margaret Forster, also in England, died in 1771 at the age of 136 years. One woman is reported to have died in Knoxville, Tennessee, in the year 1835, at the age of 154 years.

The oldest person currently, as of March 2019, is a Japanese woman honored by *The Guinness Book of Records* who, at 116, loves playing the board game *Othello*. In addition, she is up by six a.m. and enjoys studying mathematics. But Kane Tanaka has a way to go before she becomes the woman the Guinness book lists as the oldest. That achievement goes to a French woman, Jeanne Louise Calment, who lived to 122 years.

Regardless of the veracity of Shelden's examples, they serve as vehicles for transhumanists, futurists, the wealthy, and others to plan dramatically longer lives for themselves. Arrison is convinced that the goal of research on lengthening life span is to provide more healthy time, which will lead to greater wealth and prospects for happiness. To her, "No mission can be more important."[75] She points out that there is a large aging population that is wealthy. Naturally, those individuals are more than willing to spend a great deal of money to help them become and remain healthy longer. They, in concert with eager and aggressive technologists, have engaged in what

---

73   Sonia Arrison, "Living to 100 and Beyond," *Wall Street Journal*, August 27, 2011, http://online.wsj.com/news/articles/SB10001424053111190487540457652884 1080315246.

74   Samuel Sheldon Fitch, *Functions of the Lungs* (New York: S. S. Fitch & Co, 1853), 189.

75   Arrison, *100 Plus*, 198.

might be termed "The Race for the Holy Grail." The scientists recognize some hiccups, such as reticence by the Food and Drug Administration (FDA) to give needed approvals, because the agency has been set up to address specific diseases, whereas aging is a whole-body issue.[76] Another is that the disparity between the life spans of the rich and the poor is growing.[77]

Sonia Arrison contends that "scientists, backed by salespeople, mavens, and connectors, have made great headway in the march toward a brighter future, but it will be realized only when others join them in this cause. Policy makers, activists, journalists, educators, investors, philanthropists, analysts, entrepreneurs, and a whole host of others need to come together to fight for their lives. We now know that aging is plastic and that humanity's time horizons are not set in stone. Larry Ellison, Bill Gates, Peter Thiel, Jeff Bezos, Larry Page, Sergey Brin, and Paul Allen [deceased] have all recognized the wealth of opportunity in the bioinformatics revolution, but this is not enough. Other heroes must come forward—perhaps there is even one reading this sentence right now."[78]

Downsides of Life Extension

The race to develop technologies and techniques to extend lives, at least for a segment of the population, is in high gear. Some of the methods include replacement of worn-out body parts, building a human heart, and creating advanced organs—all with the help of a 3-D printer. Other scientists are working on a new goal: to promote growing old without disease. One group at Einstein College of Medicine has been involved with testing a pill intended to stave off diseases.[79] All of this meddling with natural life changes seems great, but the outcomes of

---

76 Ibid., 197.

77 Sabrina Travernise, "Disparity in Life Spans of the Rich and the Poor Is Growing," *New York Times*, http://www.nytimes.com/2016/02/13/health/disparity-in-life-spans-of-the-rich-and-the-poor-is-growing.html?_r=0.

78 Ibid., 197-198.

79 Sumathi Reddy, "Scientists' New Goal: Growing Old Without Disease," *Wall Street Journal*, March 16, 2015, http://www.wsj.com/articles/scientists-new-goal-growing-old-without-disease-1426542180.

tampering with extended lifespan are complex. For example, older women and men will look less aged. However, extended life might also lead to a higher incidence of serial monogamy, perhaps interspersed with periods of living alone potentially resulting in a decrease in meaningful relationships.

Enthusiasts argue that scientists must have *freedom to* extend life. For example, although the dramatically extended lives described by Arrison seem deceptively simple and wonderful on the surface, they are for the privileged in society. There are also pesky issues Arrison does not address. Consider retirement planning, a prime factor in aging. It is tough enough now for the largest portion of adults to come up with enough funding for retirement given revisions and deletions of pensions, changes in social security, and massive job losses. Simultaneously, dependence on favorable returns from stock market investments will increase risk.[80]

Anne Tergesen takes a blasé and uninformed attitude about society as a whole. "Planning for retirement? Look no further than your tablet or smartphone. A growing array of apps make it easier to complete many of the most basic—and most important—tasks, from saving money and creating legal documents to figuring out a second career and where to live."[81] But there is a problem. A serious one. The United States is entering into a recession, and extreme job losses are projected beginning at the end of this decade. Just what segment of the population will be able to take advantage of the apps?

One researcher has forged a consensus regarding the need for AI researchers to work on research to be sure the inevitable superintelligence does not stray from the programming and orders of its human masters. Integral to the consensus is the concept of friendly AI that would supposedly have a significant positive effect on humanity. But can we trust all in the scientific community to align themselves with the friendly approach? Research scientists and developers value

---

80   Michael Rapoport, "Longer Lives Hit Companies With Pension Plans Hard," *Wall Street Journal*, February 23, 2015, http://www.wsj.com/articles/longer-lives-hit-firms-with-pension-plans-hard-1424742593.

81   Anne Tergesen, "The Best Online Tools for Retirement Planning and Living," *Wall Street Journal*, January 19, 2015, http://www.wsj.com/articles/the-best-online-tools-for-retirement-planning-and-living-1421726470.

freedom to carry out their desires. Failure to carry out certain aspects of superintelligence herald atrocious consequences. One is singularity, when human biological enhancement or brain-computer interfaces will have progressed to the point of a greater-than-human intelligence. At that time, the world will enter the era termed *post-singularity*, at which point there will be no distinction between human and machine. AI development is a worldwide endeavor. AI creations are not just carried out by a small, well-closeted group of researchers who all think alike. Rogues abound.

6

THE CONNECTION BETWEEN HUMANOID ROBOTS, HUMANOIDS, AND SUPERINTELLIGENCE

Far from attempting to control science, few among the general public even seem to recognize just what "science" entails. Because lethal technologies seem to spring spontaneously from scientific discoveries, most people regard dangerous technology as no more than the bitter fruit of science, the real root of all evil. Far from attempting to control science, few among the general public even seem to recognize just what "science" entails. Because lethal technologies seem to spring spontaneously from scientific discoveries, most people regard dangerous technology as no more than the bitter fruit of science, the real root of all evil.

—Jacques-Yves Cousteau, Jacques Cousteau, and Susan Schiefelbein, *The Human, the Orchid, and the Octopus: Exploring and Conserving Our Natural World* (2007)

What you are about to read is not science fiction; it is the reality of a dizzying warp speed race to technologize America, its citizens, and, ultimately, the world as we know it. As expected, there are complications. Do you really want to be entertained? In a grim sort of way? The worrisome factor is that exponential growth in computing and other technologies might make it possible to build a machine more intelligent than any human. That in turn could allow the machine to have greater problem-solving and inventive skills than its human creators. If the enhanced machine breaks away from its friendly AI developers, it could create iterations of recursive self-improvement. That process in turn could potentially expand so quickly that the machine could even write its own source code to become more intelligent than humans. Isn't that an inhuman thing to do?

A moral dilemma of epic proportions is whether scientists engaged in research and development on superintelligence and, at some point, radically enhanced humanoids should have *freedom to* carry out their individual or group desired missions. The issue is that, although superintelligence and radical humanoidization may not entail the extinction of all intelligent life, those two technologies could lead to permanent destruction of a great part of humanity's potential. Why would anyone want to bring on such a calamity? Could such an event truly happen in our lifetimes?

The Controversy on Artificial Intelligence and Existential Risk

AI is the branch of computer science that includes the study and design of intelligent agents with aims to create intelligence of machines—including what are now ubiquitously termed *robots*. AI is highly technical, embracing an enormous number of tools, and it has become an essential part of the technology industry. It is big in academia, research, and business. To give an idea of the scope of its influence, the Association for the Advancement of Artificial Intelligence (AAAI) has two conferences a year on topics such as robotics and the Web. Cognitive systems, computational sustainability, knowledge-based information systems, and knowledge representation and reasoning are leading areas of research that AI embodies. Machine learning, reasoning about plans and uncertainty, and natural language processing are among basic AI tools. Artificial intelligence is not a high-tech replacement for a human brain. However, as computer development has increased so much and so fast, these machines, along with the internet and the enormous flood of data, have reinvented the way technology is built.

Are the Internet of Things and AI really great boons to society that come with many benefits? Why would we think so? Connectivity beyond traditional devices like desktop and laptop computers, smartphones, and tablets is one reason. Another is an amazingly diverse range of other devices popularly termed robots, like connections to security systems, thermostats, electronic appliances, and smart speakers in households. These and a flurry of others, such as chatbots developed relatively recently, are blithely accepted as a positive, normal part of life. But is that the life you really desire? Yes? No? Unfortunately, the future with AI is filled with unintended economic and moral consequences.

Artificial intelligence is machinery with the ability to reason and solve problems. It includes a branch of computer science for the study and design of intelligence agents and the melding of humans, robots, and machines. The aim— to create intelligence in machines and robots— is open to debate because, as might be expected, there is a wide range of opinion about how to control risks associated with AI. In 2015, Elon Musk and Sam Altman formed a nonprofit AI research company that aspires to develop and promote friendly AI. In this view of superintelligence, the agents, that is the creations', goals are aligned with ours. They would have a positive effect on humanity, benefiting society as a whole. Musk and Altman started their company because Musk stirred controversy when, during an interview at MIT, he described artificial intelligence as our "'biggest existential threat' and added 'With artificial intelligence we're summoning the demon.'"[82] He also said that he "has had longstanding concerns about the possibility that artificial intelligence could be used to create machines that might turn on humanity."[83]

However, Musk prevaricated when he proclaimed later that "there is always some risk that in actually trying to advance [friendly] AI we may create the thing that we are concerned about." Then he added that the best defense is "to empower as many people as possible to have AI. If everyone has AI powers, then there's not any one person or a small set of individuals who

---

[82] John Markoff, "Artificial-Intelligence Research Center Is Founded by Silicon Valley Investors, *New York Times*, December 11, 2015, http://www.nytimes.com/2015/12/12/science/artificial-intelligence-research-center-is-founded-by-silicon-valley-investors.html?_r=1.

[83] Ibid.

can have AI superpower."[84] That seemingly counterintuitive strategy for reducing existential risks from AI development is just one part of the widespread controversy on AI.

Thanks to the rising awareness of AI, dozens of reports on AI developments as a whole and on superintelligence in particular have emanated from academia, government, industry, and the nonprofit sector. In response, the Future of Life Institute has held two conferences on the future of artificial intelligence. One was in 2015 and the other, in 2017 and named the Asilomar Conference, was held for principled AI discussion about major change coming over unknown timescales and every segment of society. The goal of the conference was to identify a promising research direction that can help maximize the future benefits of AI. A follow-up conference was held in 2019.

The Asilomar Principles for AI Development

The large number of participants from around the world at the 2017 conference developed a list of principles for AI development called The Asilomar Principles; they were agreed upon by 90 percent of attendees.[85] The principles addressed topics ranging from research strategies to future issues, including the potential of superintelligence. The twenty-three issues were divided into three sections: research, ethics and values, and longer-term concerns. I draw upon this momentous document, initially signed by 1,273 AI/Robotics researchers and 2,541 others, to explain how humanoid development is parallel to most of the Asilomar issues.

A first thought would be that there is no relation between the Asilomar Principles and similar principles for humanoids. However, AI, which includes the study and design of intelligence agents, also includes robotics on the melding of humans, robots, and machines. The main difference is that the aim of AI is study and research of superintelligence, whereas developers

---

84 "Silicon Valley investors to bankroll artificial-intelligence center," *Seattle Times*, December 13, 2015, https://www.seattletimes.com/business/technology/silicon-valley-investors-to-bankroll-artificial-intelligence-center/.

85 Future of Life Institute, "Asilomar Principles," 2017, https://futureoflife.org/ai-principles/?cn-reloaded=1.

of humanoids have no particular long-term aims or goals. The paramount similarity is that both have potential existential risk to change the future of humanity. Following are the issues from the Asilomar Conference. Humanoids and robots are lumped together in brackets and italics as [*humanoid*].

Research Goal
- The goal of AI research should be to create not undirected intelligence, but beneficial intelligence.
- The goal of [*humanoid*] research and creations should be to create, not undirected development, but beneficial development.

Ethics and Values
- Value Alignment: Highly autonomous AI systems [*humanoids*] should be designed so that their goals and behaviors can be assured to align with human values throughout their operation.
- Human Values: AI systems [*humanoids*] should be designed and operated so as to be compatible with ideals of human dignity, rights, freedoms, and cultural diversity.
- Human Control: Humans should choose how and whether to delegate decisions to AI systems [*humanoids*] to accomplish human-chosen objectives.

Longer-term Issues
- Risks: Risks posed by AI systems [*humanoid development*], especially catastrophic or existential risks, must be subject to planning and mitigation efforts commensurate with their expected impact.
- Recursive Self-Improvement: AI systems [*humanoids*] designed to recursively self-improve or self-replicate in a manner that could lead to rapidly increasing quality or quantity must be subject to strict safety and control measures.

Superintelligence and the Friendly AI Concept

Max Tegmark, an MIT professor and president of the Future of Life Institute in the Boston area, has forged a consensus on the need for AI researchers to work on research to be sure the inevitable superintelligence does not stray from programming and orders of its human masters. His solution is to develop a positive vision for the future by building hope for the creation of a better society through AI-safety research. Integral to it is the concept of friendly AI that would have a significant positive effect on humanity.

The idea behind friendly AI is that while machine ethics outline how an AI agent should behave, friendly AI research examines how to bring about humans' desired behavior and to ensure it is properly constrained. The uncanny part is that the term *AI* or *AI agent* is used in an anthropomorphic sense, treating gods, animals or objects as if they had human qualities. This is also why the friendly AI concept conveniently applies to radically enhanced humanoids and their development. Some philosophers question whether it is possible for any truly rational agent, whether human, artificial, or in the plant or animal kingdom, to naturally be benevolent. The idea is that if they are, safeguards are not necessary and vice versa.

The *friendly AI* term emanated, at least partly, from Version 2.1 of *Transhumanist FAQ*.[86] The argument that an ever-intelligent AI will retain its ultimate goals that form a cornerstone of the friendly-AI vision was promulgated by Eliezer Yudkowsky and others in 2008. Max Tegmark concisely explains in his 2017 book *Life 3.0: Being Human in the Time of Artificial Intelligence* that "if we manage to get our self-improving AI to become friendly by learning and adopting our goals, then we're all set, because we're guaranteed that it will try its best to remain friendly forever."[87]

Tegmark continues as if writing about a human: "For an AI, the subgoal of optimizing its hardware favors both better

---

[86] Nick Bostrom, "The Transhumanist FAQ: A General Introduction," 2003, https://www.nickbostrom.com/views/transhumanist.pdf.

[87] Max Tegmark, *Life 3.0: Being Human in the Time of Artificial Intelligence* (New York: Penguin Random House, 2017), 264.

use of current resources (for sensors, actuators, computation, and so on) and acquisition of more resources. It also implies a desire for self-preservation, since destruction/shutdown would be the ultimate hardware degradation. ... In summary, we can't dismiss 'alpha-male' subgoals such as self-preservation and resource acquisition as relevant only to evolved organisms."[88]

However, although the notion of achieving security via AI safety and friendly AI sounds good on a basic level, it has flaws. Big ones. Think about humans: don't their goals and actions change numerous times during their lifetime? Wouldn't the same be true of superintelligence agents? Besides superintelligence, think about goals and values that well-intentioned humans program into humanoids. If left alone, radically enhanced ones could be further programmed, or they could self-program with choice-making abilities. They could then dismiss humans' desires for ones they prefer.

Apart from those hitches, there are pivotal all-important critical questions, such as whether we can trust research scientists and developers to align themselves with the friendly approach. Why would they want to, considering that their passions, jobs, and careers are on the line? Additionally, AI development is a worldwide endeavor, so it is likely that determined individuals and perhaps groups can believe it is their right to freely make their own changes concerning superintelligence and humanoid creations. In brief, secure does not mean *absolutely secure.*

Erik Brynjolfsson, professor at the MIT Sloan School of Management, suggested at the conference that it's crucial to develop a positive vision for the future.

> This means that we should be imagining positive futures not only for ourselves, but for society and humanity itself. In other words, we need more existential hope! Max Tegmark, also an idealist, added, if we create a more harmonious human society characterized by cooperation toward shared goals, this will improve the prospects of the AI revolution ending well. ... Do you want to own your technology or do you want your technology to own you?[89] What do you want it to mean

---

88 Ibid., 265.
89 Ibid., 333-334.

to be human in the age of AI? Please discuss all of this with those around you—it's not only an important conversation, but a fascinating one. ... Our future isn't written in stone and just waiting to happen to us—it's ours to create. Let's create an inspiring one together![90]

INTERNATIONAL POLICY GUIDELINES ON ARTIFICIAL INTELLIGENCE

By the middle of the second decade of the twenty-first century considerable activity had begun to take place about the need for oversight and regulation of the development of artificial intelligence. In early 2015, the United Nations Interregional Crime and Justice Research Institute (UNICRI) established a center on AI and robotics to "help focus expertise on Artificial Intelligence (AI) through the UN in a single agency."[91] This center, which opened in September of 2017, focuses on "understanding and addressing the risks and benefits of AI and robotics from the perspective of crime and security through awareness-raising, education, exchange of information, and harmonization of stakeholders."[92] A major emphasis is the importance of retaining human control over weapons systems and the use of force.

Another concern is international human rights. In 2017, two reports were submitted to the UN Human Rights Council (UNHCR) that discussed the implications of AI technologies for human rights. In August 2018, a report was submitted that examines the impact of AI on the information environment; it proposes a human rights framework for the design and use of technologies comprising AI by states and private actors.

The Global Legal Research Directorate in the Law Library of Congress houses the January 2019 report *Regulation of*

---

90   Ibid., 335.

91   AI Policy – United Nations, Future of Life Institute, https://futureoflife.org/ai-policy-united-nations/ (last visited January 4, 2019), archived at https://perma.cc/9CZ2-ETNX.

92   Centre on Artificial Intelligence and Robotics, UNICRI, para. 5, n.d., http://www.unicri.it/topics/ai_robotics/centre/ (last visited Dec. 10, 2018), archived at https://perma.cc/475Y-ZZ6N.

Artificial Intelligence in Selected Jurisdictions. That document contains a comparative summary of the emerging regulatory and policy landscape surrounding artificial intelligence in jurisdictions around the world and in the European Union. It contains the results of a survey of international organizations that describe the approach United Nations (UN) agencies and regional organizations have taken toward AI.

The Organization for Economic Co-operation and Development (OECD) is made up of thirty-six countries. That organization adopted a set of Principles on Artificial Intelligence in May 2019 to promote AI that is innovative and trustworthy, and that respects human rights and democratic values. The report states that these Principles "are practical and flexible enough to stand the test of time in a rapidly evolving field. They complement existing OECD standards in areas such as privacy, digital security risk management, and responsible business conduct" and that they "are the first such principles signed by governments beyond OECD members."[93]

The OECD Council Recommendation on Artificial Intelligence identified five complementary values-based principles for the responsible stewardship of trustworthy AI: (1) AI should benefit people and the planet by driving inclusive growth, sustainable development, and well-being, (2) AI systems should be designed in a way that respects the rule of law, human rights, democratic values, and diversity, and they should include appropriate safeguards—for example, enabling human intervention where necessary–to ensure a fair and just society, (3) there should be transparency and responsible disclosure around AI systems to ensure that people understand AI-based outcomes and can challenge them, (4) AI systems must function in a robust, secure, and safe way throughout their life cycles and potential risks should be continually assessed and managed. (5) Organisations and individuals developing, deploying or operating AI systems should be held accountable for their proper functioning in line with the above principles.

These guidelines are significant because the principles have managed to unite nations at a time when there is little international cooperation, and to reinforce the importance of

---

[93] OECD, "OECD Principles on Artificial Intelligence," May 22, 2019, para. 1, https://www.oecd.org/going-digital/ai/principles/.

values in AI development that promote AI that is innovative and trustworthy and respects human rights and democratic values.

Stanford University launched the Stanford Institute for Human-Centered Artificial Intelligence (HAI) on March 18, 2019, with the mission of advancing artificial intelligence research, education, policy, and practice to improve the human condition. The objective is "to serve as a rallying point and catalyst for interdisciplinary collaboration to realize a shared dream of a better future for all of humanity." The belief is that the Institute, with its unique cross-campus diversity of thought and its collaboration with industry and with policy makers, has the potential to help mitigate the "challenges and disruptions that societies around the world will need to be prepared to confront."[94]

The Center for Governance of AI, housed at the Future of Humanity Institute at the University of Oxford, strives to help humanity capture benefits and mitigate risks of AI. The focus is on the political challenges arising from transformative AI, systems whose long-term impacts may be as profound as those of the Industrial Revolution. The Center seeks to guide development of AI by conducting research on important and neglected issues of AI governance. For example, a report, *The Malicious Use of Artificial Intelligence: Forecasting, Prevention, and Mitigation,* explores possible risks to security posed by malicious applications of AI, including in digital, physical, and political domains. The report lays out a research agenda for further work addressing such risks. Not only is artificial intelligence a hot topic, but the funding is rolling in to fund and escalate this awesome technology that, according to many, will change the world in positive and productive ways. Oxford University made an announcement that its biggest ever cash donation will be directed partly toward the establishment of the Ethics in AI Institute, scheduled to open in 2024.

The National Science Foundation, an independent U.S. federal government agency, supports "fundamental research to bring technologies to maturity, thereby enhancing the

---

94 Amy Adams, "Stanford University launches the Institute for Human-Centered Artificial Intelligence," March 18, 2019, https://news.stanford.edu/2019/03/18/stanford_university_launches_human-centered_ai/.

lives of all Americans.... The effects of AI will be profound. To stay competitive, all companies will, to some extent, have to become AI companies. We are striving to create AI that works for them, and for all Americans."[95] The NSF has an annual budget of $8.1 billion (FY 2019) of which over $100 million is directed toward AI research.[96] On September 11, 2019, there was an announcement of a nearly $1 billion federal commitment toward artificial-intelligence research. The announcement drew a mixed response from executives, some of whom claimed that $1 billion for AI research isn't enough.[97]

Prognostications on Our AI Future

As I conducted my research, I realized the public has little knowledge of, or wherewithal to demand, *freedom from* risky AI developments that loom on the near horizon. At that point, I became upset, because some things are just morally wrong. That was when I decided to take on the task of writing about who and what is behind such developments. Think about this. In the words of Max Tegmark, "Will we control intelligent machines or will they control us? Will intelligent machines replace us, coexist with us, or merge with us? What will it mean to be human in the age of artificial intelligence? What would you like it to mean, and how can we make the future that way?"[98]

Warp speed artificial intelligence creations have led to a number of books on superintelligence and other aspects of AI in the mid-term and far-out future. In addition to Max Tegmark's book *Life 3.0*, let's take advantage of several well-known authors on technology for their views on the impact AI could have on humankind. There is Robin Hanson, author of

---

95   National Science Foundation, "Statement on Artificial Intelligence for American Industry," press release 18-005, May 10, 2018, https://www.nsf.gov/news/news_summ.jsp?cntn_id=245418.

96   National Science Foundation, "About the National Science Foundation," n.d., https://www.nsf.gov/about/.

97   Artificial Intelligence Daily, "Feds to Spend $1 Billion on AI Research," WSJ Pro AI, September 11, 2019, access@interactive.wsj.com.

98   Tegmark, *Life 3.0*, 38.

the 2016 book *The Age of Em: Work, Love and Life When Robots Rule the Earth*, who projected that one day, the first truly smart robots, which he calls Ems, will be brain-emulation robots that may rule the world. The concept of Ems reveals how strange humans' descendants may be if there are no controls and bans on humanoid, superintelligence, and other AI development. It also demands critical thinking about common assumptions regarding humanity's moral progress and what we hold dear.

Futurist Yuval Noah Harari, an ancient history specialist, did not participate at the Asilomar conference. Instead, he followed in the footsteps of other prophet-like long-term authors when he opined in his 2018 book, *21 Lessons for the 21st Century*, that by 2050, unenhanced humans will have become completely useless. Harari is convinced that, simultaneously, AI creations will take the place of humans. That setting will leave them with vast leisure to happily enjoy their lives.

On a seemingly positive and AI-friendly side is Nick Bostrom. In his profoundly ambitious 2014 book *Superintelligence: Path, Dangers, Strategies,* he wrestles with how to promote "best practices" among AI researchers to ramp up safety if and when the prospect of machine intelligence begins to look imminent. His prescription is to push forward on singularity research and to maintain commonsense and good-humored decency, even in the teeth of this inhuman conundrum.

Stuart Russell, a professor of computer science and holder of the Smith-Zadeh Chair in Engineering at the University of California, Berkley, lays out an approach in his 2019 book, *Human Compatible: Artificial Intelligence and the Problem of Control*, that will enable humans to coexist successfully with increasingly intelligent machines. Along the lines of Max Tegmark, he suggests that we can rebuild AI on a new foundation, in which machines are designed to be inherently uncertain about human preferences they are required to satisfy. Such machines would be humble, altruistic, and committed to pursuing our objectives, not theirs. Russell makes no mention of issues regarding existential risk.

On a technical level, Adam Piore in his 2017 book, *The Body Builders: Inside the Science of the Engineered Human*, explores the current revolution in human augmentation, taking us into the field of bioengineering and introducing us to the people at its center. Piore argues that the new scientific frontier is the human body and that the greatest engineers of our generations have turned their sights inward and are beginning to revolutionize mankind. His viewpoint, akin to that of transhumanists, is that this revolution is helping humankind to triumph over the limitations and constraints long accepted as an inevitable part of being human. Like Russell, Piore does not discuss existential risk.

Amir Husain, an acclaimed technologist and inventor, argues in his 2017 book *The Sentient Machine: The Coming Age of Artificial Intelligence* that it is wrong to start contemplation of AI from a place of existential fear rather than from one of opportunity. Rather, Husain suggests that we ought to take an encouraging view of how we can live amid the coming age of sentient machines and artificial intelligence, because we will still have purpose and will be creators of AI, our greatest creation of all. His goal is to remind his fellow engineers and scientists that they are misguided in promoting bans on technologies because doing so would only slow the progress of so many opportunities for making our world a more just and equitable place.

On the practical current public policy side is Martin Ford, a futurist who focuses on the impact of AI and robotics on society and the economy. He presents a strong argument in his 2016 book, *Rise of the Robots: Technology and the Threat of a Jobless Future*, that advances in robotics and artificial intelligence would eventually make a large fraction of the human workforce obsolete. As I do in this book, he argues that the growth of automation even threatens many highly educated people, like lawyers, radiologists, and software designers.

Ford's thoughts cannot be taken lightly. They demand serious consideration of the existential risk involved—a risk that cannot be undone, one that poses permanent and significant negative consequences for humanity. The hardcore reason is that creation of friendly AI and superintelligence is a risk-taking venture because humans will depend on the friendliness both of creators of superintelligence and of the machine creations. Now, a bottom-line, critical issue: Most futurists, and a large portion of the public and scientists, *presume* that radical AI development is inevitable, just as change is inevitable. But is it? I argue no, and later in this book I will reveal how controls can be justified and utilized.

Post-Singularity: An Irresponsible Option for Our Future

Failure to control certain aspects of technological development presage atrocious unintended consequences. Singularity is a result of superintelligence. The term *technological singularity* is also used to mean a hypothetical moment when artificial intelligence, human biological enhancement, or brain-computer interfaces will have progressed to the point of a greater-than-human intelligence that will radically change civilization, and perhaps even human nature. At that time, the world will enter the era termed post-singularity, at which point there will be no distinction between human and machine.[99] Some equate singularity with technomancy in which use of technology leads to magical abilities or powers.[100] The result of unfettered AI development would be radically enhanced humanoids, those that have advanced capacities exceeding those of present humans. At that point, humanoids would no longer be unambiguously human by our current standards.

Some critics shrug off the singularity issue as being a philosophical matter. Others contend that there is no need to decide whether a machine can think, but only to decide whether if a machine can think as intelligently as humans can. Still other critics, such as Stephen Hawking, Stuart Russell, Max Tegmark, and Frank Wilczek, have argued that little serious research has been conducted outside of institutions devoted to examining what is the best or worst thing that could happen to humanity. These include:

- the Cambridge Centre for the Study of Existential Risk,
- the Future of Humanity Institute, and
- the Machine Intelligence Research Institute.

---

99 Ray Kurzweil, *The Singularity Is Near: When Humans Transcend Biology* (New York: Penguin Books, 2005), 9.

100 The term *Technomancy* is in Arthur C. Clark's third law of prediction in his book *Profiles of the Future: An Inquiry into the Limits of the Possible*. Originally published in 1962 and updated in 1984, it is available in paperback (London: Orion Publishing Group, 2000).

Hawking's opinion was that developing AI could be humanity's most detrimental mistake.[101]

The most alarming aspect of singularity rests on the notion that such a point could happen suddenly, out of the hands of superintelligence developers. On the "Yes, it can and likely will happen" side is the well-known technologist Ray Kurzweil, who wrote, "With reverse engineering of the human brain we will be able to apply the parallel, self-organizing, chaotic algorithms of human intelligence to enormously powerful substrates. This intelligence will then be in a position to improve its own design, both hardware and software, in a rapidly accelerating iterative process."[102] He strongly argues that technological singularity can occur by 2045, about two and a half decades hence. Writing in contrast, and much more disconcerting, is the well-known science fiction writer Vernor Vinge, who has postulated that singularity can take place sometime before 2030, about ten years from now. The dilemma he poses is that we just don't know when and if singularity might strike, so all the more reason for an abundance of caution—and precautions.

Kurzweil maintains that humans are just a facilitator for what flows through us, and therefore we shouldn't worry about selfhood and identity. To him, we will be living in a much better world when we create conscious, independent artificial intelligences. These machines will be conscious because they convincingly tell us they are. We should create them to augment our intelligence that, in turn, will quickly solve minor problems such as war, death, and material scarcity here on earth. Then we can head out and colonize the universe. Kurzweil concludes, "Waking up the universe, and then intelligently deciding its fate by infusing it with our human intelligence in its non-biological form, is our destiny."[103] I, however, believe that most people would disagree.

Here is a dilemma. Are ordinary citizens really prepared and of sound enough mind to decide on our destiny? Is our

---

101 George Dvorsky, "Stephen Hawking Says A.I. Could Be Our 'Worst Mistake in History,'" *Futurism*, May 2, 2014, http://io9.com/stephen-hawking-says-a-i-could-be-our-worst-mistake-in-1570963874.

102 Ibid.

103 Ray Kurzweil, *How to Create a Mind: The Secret of Human Thought Revealed* (New York: Viking Penguin, 2012), 282.

populace ready to let the system potentially bring a sort of Armageddon on the world's peoples?
- Will the system fail our citizens?
- Do citizens really care that America is essentially a plutocracy with extreme power over our republican democracy?
- Will our government separate itself from the technocracy and plutocracy?

7

THE PUBLIC'S RIGHTS VERSUS THE PLUTOCRACY, TECHNOCRACY, AND GOVERNMENT

Science is a magnificent force, but it is not a teacher of morals. It can perfect machinery, but it adds no moral restraints to protect society from the misuse of the machine. It can also build gigantic intellectual ships, but it constructs no moral rudders for the control of the human vessel. It not only fails to supply the spiritual element needed but some of its unproven hypotheses rob the ship of its compass and thus endanger its cargo.

—William Jennings Bryan, *Scopes Monkey Trial Summation* (1925)

The emphasis of this book at this point shifts from technical aspects to the actors who will dramatically affect our destinies, particularly in the United States, but also in the world as a whole. Some key questions:
- Will the republican democracy experiment by our founders run its course?
- Do you think it will be possible to rectify our democratic system for contemporary and future generations?
- How about our dysfunctional congressional system? Can it be rectified?
- Will Congress be able to overcome the technocracy and plutocracy to save humanity from AI risks in the form of superintelligence and radically enhanced humanoids?
- What is required to protect humanity as we know it?
- Closer to home, will the media that provides us with the news of the world and the home front be able to survive beyond a decade or so from now?

Enter the Triumvirate

What we do feel, but cannot articulate, is that technical experts are a meritocracy in the social and political system. The higher-ups are the technocrats, members of the powerful technological elite who have a great deal of influence in politics and industry. This grouping, which advocates for the supremacy of technical experts, has inordinate power. Along with the wealthy, they are directly in bed with our government. This loose triumvirate—the technological elite, the wealthy, and the government—are the ones that have dominance over technological policies. For this book, the concern is the triumvirate's control over existential risks to humans arising from superintelligence and radically endowed humanoids.

Let me hasten to say that the big three is not a cabal, which is a small group of people involved in secret plans to get political power. Additionally, there is no conspiracy among the triumvirate. However, collectively, their power and ideologies give them a strong voice on consequential scientific issues—the relevant one here being whether to regulate or ban certain types of research and the use of technologies that can lead to permanent consequences for humanity. There are many issues and bones of contention. We feel a nagging fear that, despite great hype and spin about how technology will save us all, it may in fact be our undoing. Deep down, we lack faith and trust in government.

The Plutocracy Is at Odds with America's Republican-Style Democracy Republicanism

The sad truth is that America has rapidly become a society dominated by the plutocrats—defined as those who are powerful because of their wealth—a diverse group intent on influencing government. The plutocracy is also known as *plutonomy* (from the Greek *ploutos*, meaning wealth + nomos meaning law— the science of production and distribution of wealth) or *plutarchy*— these are terms generally used in a pejorative sense. In daily life they refer to any form of government in which the wealthy, particularly the überwealthy, exercise the preponderance of power, whether directly or indirectly. Additional players tethered to it are strong and powerful company CEOs and the military–industrial complex with its associated interests and lobbyists. In brief, America is a plutocracy that is narrowing and becoming more demanding. Many say America already is an oligarchy, a form of government in which only a small group of people hold all the power.[104]

Plutocracy is not something new in America. Its antecedents lie in the period between the Civil War and the Great Depression, when a handful of very wealthy heads of large corporations with monopolistic or near monopolistic control increasingly exerted influence over politics and public opinion. The enormity of direct and indirect control by the powerful wealthy in that period is emphasized by Nobel Prize–winning economist Joseph E. Stiglitz in his enduring 2011 *Vanity Fair* article, "Of the 1%, by the 1%, for the 1%." The situation is much the same today. To Stiglitz, although the current plutocracy is not a deliberate conspiracy to grab power, it is all about the erosion of values that have been the bedrock of our nation: "identity, fair play, equality of opportunity, and sense of community."[105]

To Noam Chomsky, the belief that America has a citizen-run democracy is a myth. His reasoning is that many high-level elected officials are wealthy and powerful enough to have their own agenda and interests, impervious to those beneath them. To Chomsky, without question, lobbyists at the higher levels in the American system are minions of the wealthy and powerful and are therefore essentially agents of the plutocracy. Paul Krugman has provided the same message, observing, "We have a society in which money is increasingly concentrated in the hands of a few people, and in which that concentration of income and wealth threatens to make us a democracy in name only."[106]

Our country is touted as being an exceptional one because it has a republican-style democracy of the people, by the people,

---

104 Some will argue that America is not a plutocracy. That issue is discussed at length in William Domhoff, *Who Rules America? Power, Politics, and Social Change* (New York: McGraw-Hill, 2005) and in a companion book by Domhoff, *Who Rules America? Challenges to Corporate Class Dominance* (New York: McGraw-Hill, 2009).

105 Joseph E. Stiglitz, "Of the 1%, By the 1%, For the 1%," May 2011, para. 11, https://www.vanityfair.com/news/2011/05/top-one-percent-201105.

106 Paul Krugman, "Democracy and American Oligarchs," *Seattle Times*, November 5, 2011.

for the people. In brief, we *supposedly* have a system controlled by the people and managed by a beneficent government for the good of all citizens. However, that belief, widely taught in schools, has in actuality morphed into a situation where raw power is wielded by an increasingly self-serving state that justifies secrecy and duplicity by saying they are necessary to protect Americans' security. America is, in this sense, unexceptional.

Wealthy individuals and organizations can exert influence on the political arena by either financing their own political campaigns or affiliating with other wealthy persons and organizations. That power compromises the American democratic system, because a core basis of the system is that Congress and other elected persons should have the ability to freely advocate for policies their constituents want. Calamitously, the electoral college makes nightmares possible.

An example of concern is that, through monetary power, the United States has in a few decades lost the foundation of democracy. The *Citizens United* ruling by the Supreme Court is a case in point. It removed the previous ban on corporations and organizations using their treasury funds for direct advocacy. These groups were then freed to expressly endorse, or call to vote for or against, specific candidates, actions that were previously prohibited.[107] In effect, this ruling was a death knell to those not born to the wealthy and elite classes.

Capitalism Is Sacred in America

Capitalism is a sacred value in America. At least it was. The role of business is supposedly to provide products and services that make people's lives better and thus make the populace happier. That is close to the vision held in the late nineteenth and early twentieth centuries during the early Progressive Era. Such thinking was the basis for political struggles about socialism and communism in pre-WWII America. At that time, lifetime

---

107 Campaign Legal Center, "A Guide to the Current Rules for Federal Elections: What Changed in the 2010 Election Cycle," November 7, 2010, https://campaignlegal.org/press-releases/guide-current-rules-federal-elections-what-changed-2010-election-cycle.

employment and pensions were considered commonplace, an appropriate lifestyle for the future.

By 1955, roughly a third of American workers were union members. Our capitalistic system actually did work along the Progressive Era lines until the 1960s. The author of a 1955 *Fortune* magazine essay, "How top executives live," compared the vast mansions, armies of servants, and huge yachts of the 1920s with the modest lifestyles of typical executives in 1955. Those executives lived in smallish suburban houses. They relied on part-time help and skippered their own relatively small boats.[108] The point is, we need heroes and role models from that era to fathom how far off the beam our country has gone. John Bogle, founder of the investment giant Vanguard, provides a perfect example of how his old-fashioned values set was, and still is, widely admired. He passed away at age eighty-nine in 2019.

At first Bogle was reviled and ridiculed, but for decades his mix of pragmatism and idealism enabled millions of ordinary Americans to build wealth to buy a home, pay for college, and retire comfortably. "Jack could have been a multibillionaire on par with Gates and Buffett," said William Bernstein, an Oregon investment manager and author of twelve books on finance and history. Instead, Bogle turned his company into one owned by its mutual funds, and in turn, their investors. According to Bernstein, the company "exists to provide its customers the lowest price. He [Bogle] basically chose to forgo an enormous fortune to do something right for millions of people. I don't know any other story like this in American business history."[109] Bogle was fiercely competitive when it counted. But he was willing to compromise when necessary. He was also loving, sentimental, kind, charitable, and courageous. Most important, he taught us humility, ethics, and simplicity, values that are now mostly seen as curious relics of a past era.

Equally poignant is Stiglitz's classic 2012 book *The Price of Inequality*, which focuses on values that include equality,

---

108   Paul Krugman, "The Upside of the 'Twinkie Era,'" Opinion, *Seattle Times*, November 20, 2012.

109   Art Carey and Erin Arvedlund, "John Bogle, who founded Vanguard and revolutionized retirement savings, dies at 89," *Philadelphia Inquirer*, January 16, 2019, http://www.philly.com/business/john-bogle-dead-vanguard-obituary-20190116.html.

fairness, and fair play in the economic system. Stiglitz argued that predatory lending led to the Great Recession (December 2007–June 2009). He noted:

> ...what is remarkable is how few seemed—and still seem—to feel guilty, and how few were the whistleblowers. Something has happened to our sense of values, when the end of making more money justifies the means, which in the U.S. subprime crisis meant exploiting the poorest and least-educated among us. ... Much of what has gone on can only be described by the words 'moral deprivation.' Something wrong happened to the moral compass of so many people working in the financial sector and elsewhere. When the norms of a society change in a way that so many have lost their moral compass, it says something significant about the society.[110]

As the twenty-first century has unfolded, unassuming top executives and concern for the well-being of all society have largely disappeared. Big business is now increasingly vilified by the American public, and for good reason. For example, a Rasmussen poll conducted in 2016 revealed that 68 percent of voters "believe government and big business work together against the rest of us."[111] Businesses feeding at the public trough and being able to gain favors in securing government largesse such as grants, loans, favorable regulations, and no-bid contracts has acquired the moniker *corporate cronyism*. Worst of all, there is little or no oversight.[112] This cronyism is viable because the system is so big. It is so complicated that it is out of control and, for all purposes, operating on autopilot. Most of the blame falls on the speed of change in all aspects of America, from the cultural to the technological realms.

---

110   Joseph E. Stiglitz, *The Price of Inequality* (New York: W.W. Norton & Company, 2012), xvii.

111   Rasmussen Report, "68% Think Government, Big Business Often Work Together Against Consumers, Investors," July 2016, http://www.rasmussenreports.com/public_content/politics/general_politics/july_2016/most_think_government_big_business_work_together_against_america.

112   The wealthy libertarian Charles G. Koch argues, "When businesses feed at the federal trough, they threaten public support for business and free markets." See "Charles G. Koch: Corporate Cronyism Harms America," *Wall Street Journal*, September 9, 2012.

Equally, a major slice of blame goes to gridlock in Congress and its systemic continual rancor over values-laden issues and ideological viewpoints.

As if all of this were not enough, tech giants Amazon, Apple, Facebook, and Google have built themselves into some of the largest players in terms of influence on, and open access to, industry. The *New York Times* reported, "Faced with the growing possibility of antitrust actions and legislation to curb their power; four of the biggest technology companies are amassing an army of lobbyists as they prepare for what could be an epic fight over their futures. The four companies spent a combined $55 million on lobbying in 2018 and are spending at a higher rate in 2019."[113]

A further overriding problem in America's capitalist-oriented system is that medium and long-term planning for governmental policies are not favored. The mishmash of policies and legislation between the states and federal government makes finding consensus difficult, domestically and globally. Thus, there is no vision guiding long-term goals concerning technological advances. The point is that, if citizens are to be in charge of development of technologies that exhibit existential risk, there must be a strong government devoted to regulation of the wealth and power of the elite. That is not now the state of affairs.

The Congressional System Has Lost Its Bearings

Does our government system, as it stands currently, have the best interests of the 99 percent at heart? Sadly, there is reason to doubt that that is the case. It is inconceivable that there is no widespread angry rancor to modernize our congressional system, despite a lot of griping. D. Elisabeth Glassco pointed out that one reason for the lack of action is a largely dysfunctional Congress. Another is that the government is beholden to its corporate and wealthy donors. A further one is the lucrative

---

[113] Cecilia Kang and Kenneth P. Vogel, "Tech Giants Amass a Lobbying Army for an Epic Washington Battle," *New York Times*, June 5, 2019, https://www.nytimes.com/2019/06/05/us/politics/amazon-apple-facebook-google-lobbying.html.

prospect of finding employment after leaving government, through those influential donors—the revolving door. Glassco also noted that Congress had reached a historically low approval rating falling to around 9% percent in November of 2013, amid the litany of large-scale scandals, coverups, and misdealing's by top government officials coupled with the U.S. government shutdown.[114] The rating hovered at around 17 percent during 2018, well below the overall average of 30 percent since Gallup began tracking the public's views of Congress in 1974.

Civility, an indispensable value and mindset about how to act, is not yet dead, but it is now generally considered old-fashioned. However, incivility is not just a new standard. American history is replete with incivility, which is as old as politics itself. In fact, the middle of the nineteenth century was a time of unstinting incivility as sectional strife generated by slavery began the long march to the Civil War.[115] Interestingly, the congressional chambers were turned into a sort of armed squabbles where both pro-slavery and anti-slavery congressmen carried on their persons a host of weaponry while attending debates. As one senator observed of the times, "[T]he only congressmen not carrying a knife or a revolver were those carrying *two* revolvers!"[116]

Irrefutably, as is documented in the book *Civility and Democracy in America: A Reasonable Understanding* by Cornell Clayton, director of the Thomas S. Foley Institute for Public Policy and Public Service and Richard Edgar, assistant director of the same Institute, civility is at its lowest point since Civil War times. The authors argue, "Political leaders and their partisan followers do not necessarily need to agree, but the ability to consider different viewpoints and to respect one another as common citizens is a crucial aspect of democratic

---

114 Frank Newport, "Congressional Approval Sinks to Record Low," November 12, 2013, http://www.gallup.com/poll/165809/congressional-approval-sinks-record-low.aspx.

115 D. Elisabeth Glassco, "Historical Incivility as a Foundation of America's Democracy," *Medium Politics*, August 18, 2018, https://medium.com/@deglassco/historical-incivility-as-a-foundation-of-americas-democracy-pt-2-791064975c4b.

116 Don E. Fehrenbacher, *The South and Three Sectional Crises* (Baton Rouge: Louisiana State University Press, 1980), 62; Paul F. Boller, *Congressional Anecdotes* (New York: Oxford University Press, 1992).

governance."[117] The truth is civility, like so many quaint practices of earlier years, is inconvenient for the emergent technocracy, the plutocracy, and many leaders of our government.

This new era's reality of operating at warp speed has led to a decline in reading and contemplating printed media like newspapers. Instead, the public gathers information through other means, mainly their gadgets. As a result, critical thinking applied by reading is declining rapidly. The public is becoming what might be termed "factoidized" through news bites and entertaining breaking news. Those who do venture beyond sound bites might have pause to wonder whether the Generation Z and Post-Zers will pay homage to our American republican-style democracy. This new generation has reason to be bewildered, considering the antics in the political arena, such as brinkmanship regarding the budget. And then there is the outlandish increase of the national debt, and a president who shut down much of government to satisfy his egotistical desire for a $5.6 million wall across the border with Mexico. Without a doubt, this new reality is a major factor in why citizens, and most of the world, are developing mistrust of U.S.-style democracy.

Is America's republican democracy actually in danger due to the demise of our once free press system? The alarming answer is yes, as the *Seattle Times* reported on December 22, 2019. Polls increasingly show that young people question whether our democracy is the best form of government and whether capitalism is the best economic system for an economy. Frank Blethen, publisher of the *Seattle Times*, points out in a 2019 article titled "Save the Free Press initiative: A public service of the Seattle Times" that "Sixty-nine years ago, our free press system—the foundation of America's democracy—was strong, stable and robust,"

Blethen pointed out a litany of historical reasons for why newspapers are essential if our democracy is to continue. One was described by noted journalist Walter Lippman in 1950, who Blethen quotes as noting: "'The secret of a truly free press is it should consist of many newspapers, decentralized in their ownership and their management, and dependent for their

---

117   Cornell Clayton and Richard Elgar, *Civility and Democracy in America: A Reasonable Understanding* (Pullman, WA: Washington State University Press, 2012), xi.

support upon the communities where they are written, where they are edited and where they are read.'"

Blethen adds that Thomas Jefferson wrote in 1787 that if it were up to him to "'decide whether we should have a government without newspapers or newspapers without government, I should not hesitate a moment to prefer the latter.'"

Blethen states, "Today would be Walter Lippmann and Thomas Jefferson's worst nightmare: the demise of our once robust free press system and the threat to the very survival of our democracy.

"History tells us that few democracies survive beyond 200 years, and almost all democracies that die, die from within—not from without."

In a similarly titled 2019 article, "Save the Free Press initiative: A message from Publisher Frank Blethen," Blethen observes that the free press system in our nation is "literally on life support as it suffers through the final stages of disinvestment and destruction by a handful of non-democratic fiscal oligarchies (hedge funds and distressed-asset players) who control most of the country's newspapers. These are the bottom-feeders at the end of a four-decade period of consolidation and lost local control—something that we should never have tolerated because it put our democracy in peril."

"Survival of our country's free press is a crisis which requires great urgency. Without reform and re-creation of a modern free press system, it is hard to envision our democracy surviving for another decade."[118]

But at the same time, there was a rare bit of good news for an industry that has struggled to make ends meet. Tucked into a last-minute $1.4 trillion spending bill signed on December 20, 2019, was legislation known as the Save the Community Newspaper Act (SCNA). That new law, relevant to publications that are privately owned and operate mainly within a single state, allows affected newspapers to defer some of the contributions they make into their employee pension funds. The method is somewhat like homeowners changing the term of a home mortgage from fifteen to thirty years to reduce the amount of each monthly payment. That *temporary* relief for as many as seventeen newspapers will offer more time

---

118   Frank Blethen, *Seattle Times*, prepared on December 20, 2019 and published on December 22, 2019, A11.

to replace ad revenue with other revenue sources. "Combined with proceeds from recent property sales," the *Seattle Times* reported, "that effectively gives The Times another decade to complete the ongoing shift to new revenue sources, including digital subscriptions and community-funded coverage."[119]

Now critical questions arise: will newspapers be able to survive beyond a decade or so from now? Will it be possible to rectify our democratic system for contemporary and future generations, or has the republican democracy experiment by our founders run its course? Will our dysfunctional government dissolve into a full-blown plutocracy? More at home, will our congressional system be able to overcome the technocracy and plutocracy to save humanity from AI risks in the form of superintelligence and radically enhanced humanoids? Bottom line: what can be done to repair our governmental system emphasizing a country of the people, by the people, for the people?

The Triumvirate's Hold on the System

Jeffrey D. Sachs pointed out something few people associate with politics and the wellbeing of America: happiness. Sachs, the author of "Restoring American Happiness" in the *World Happiness Report 2017*, summed the situation up by explaining, "The predominant political discourse in the United States is aimed at raising economic growth, with the goal of restoring the American Dream and the happiness that is supposed to accompany it. But the data show conclusively that this is the wrong approach. The United States can and should raise happiness by addressing America's multi-faceted social crisis—rising inequality, corruption, isolation, and distrust—rather than focusing exclusively or even mainly on economic growth, especially since the concrete proposals along these lines would exacerbate rather than ameliorate the deepening social crisis."[120]

Chrystia Freeland, in her book *Plutocrats: The Rise of the New Global Super-Rich and the Fall of Everyone Else*, is

---

[119] Paul Roberts, *Seattle Times* business reporter, "Last-minute federal legislation brings pension relief, and much-needed 'runway,' for community newspapers," *Seattle Times*, December 20, 2019.

[120] Jeffrey D. Sachs, *World Happiness Report 2017*, produced by the United Nations Sustainable Development Solutions Network in partnership with the Ernesto Illy Foundation, 2017, https://worldhappiness.report/ed/2017/.

instructive about what will happen given that everyone is on the payroll of the plutocrats in one way or another. However, as she sees it, while these elites like to think of themselves as acting in the collective interest, actually they are only concerned about their vested interests.[121] That is a tragic appraisal, but not a new one. Adam Smith, often considered the father of modern economics, first used the metaphor of the "invisible hand" in his 1776 book *The Wealth of Nations* in reference to merchants and entrepreneurs. Smith wrote, "He intends only his own security; and by directing that industry in such a manner as its produce may be of the greatest value, he intends only his own gain, and he is in this, as in many other cases, led by an invisible hand to promote an end which was no part of his intention."[122]

The triumvirate is one component of the system we all face in which many essentially give up and say, "You can't beat the system," meaning in this case that one might as well just give up and let the system run its course. The problem is that the issue is not about wealth per se or about how multimillionaires and billionaires spend their money. Rather, it's about their power to control how the system works. An overriding concern, as is directly or indirectly expressed in writings about negative utopias, is that at some point, the plutocracy will gain so much power that America will indeed transition into a malevolent oligarchy. An alternative is to only *hope* (or as some folks prefer, *pray!*) that American-style democracy will not fail them.

Several noted authors have critically evaluated the state of American wealth and leadership. Paul Krugman, noted economist and Nobel laureate, charges that "money is the glue of movement conservatism, which is largely financed by a handful of extremely wealthy individuals and a number of major corporations, all of whom stand to gain from increased inequality, an end to progressive taxation, and a rollback of the welfare state—in short, from the reversal of the New Deal. And turning the clock back on economic policies that limit inequality is, at its core, what movement conservatism is all about."[123] He

---

121 Chrystia Freeland, *Plutocrats: The Rise of the New Global Super Rich and the Fall of Everyone Else* (New York: The Penguin Press, 2012).

122 Adam Smith uses the metaphor in Book IV, Chapter II, paragraph IX of *The Wealth of Nations*, first published on March 9, 1776.

123 Paul Krugman, *The Conscience of a Liberal* (New York: W.W. Norton & Company, 2007), 10.

continues, "Because movement conservatism is ultimately about rolling back policies that hurt a narrow, wealthy elite, it's fundamentally antidemocratic."[124]

Columnist Peggy Noonan provided an insight into the mindset of the elites, writing, "An odd thing, in my observation, is that deep down the elite themselves also think the game is rigged. They don't disagree, and they don't like what they see—corruption, shallowness and selfishness in the systems all around them. Their odd anguish is that they have no faith the American people can—or will—do anything to turn it around. They see the American voter as distracted, poorly educated, subject to emotional and personality-driven political adventures."[125] She continues, "Both sides, the elites and the non-elites, sense that things are stuck. The people hate the elites, which is not very new, and very American. The elites have no faith in the people which, actually, is new. Everything is stasis."[126]

In sum, America's elites, like elites in much of the developed world, have willfully blinded themselves to economic inequality. They consider themselves meritocratic with good reason, for they have worked hard to adapt to the needs of a technology-driven, globalized economy. They embrace competition, because most of the time, they win. The dilemma is that the triumvirate will win if there are no regulations and bans on radically enhanced humanoids and superintelligence. The answer is the government must separate itself from the technocracy and plutocracy because for the elite, freedom means *freedom to do* whatever they want to do. To them the proper refrain, as Owen Wister wrote in his 1902 immortal book *The Virginian: A Horseman of the Plains* is "may the best man win."[127]

The triumvirate can also win through diplomatic sleight of hand, because government policy and influence by the

---

124 Ibid., 11.

125 Peggy Noonan, "America Is So in Play," *Wall Street Journal*, August 27, 2015, http://www.wsj.com/articles/america-is-so-in-play-1440715262.

126 Ibid.

127 "It was through the Declaration of Independence that we Americans acknowledged the eternal inequality of man. ... 'Let the best man win, whoever he is.' Let the best man win! That is America's word. That is true democracy. And true democracy and true aristocracy are one and the same thing. If anybody cannot see this, so much the worse for his eyesight." Owen Wister, *The Virginian: A Horseman of the Plains* (New York: Macmillan, 1902), 108.

plutocracy and technocrats is out of control. If no publicly debated decision is made, then the organizational assumption known as "catch-22," the unwritten law of informal power that exempts the organization from responsibility and accountability, will take hold. That enigma plays out in the final episode in Joseph Heller's 1961 novel *Catch 22*. An old woman recounts an act of violence by soldiers: "Catch-22 says they have a right to do anything we can't stop them from doing."[128]

Isn't there a safe harbor in the legal system to protect the public from harm when scientific investigation has found plausible risk that a threat from serious or irreversible damage to the environment or health exists? Short answer, yes. Have you ever heard about the Precautionary Principle that evolved out of the German socio-legal tradition, created in the heyday of democratic socialism in the 1930s? How about the United States as the logical site for a working group in international proceedings on the Principles? That's not likely to happen now, given a lack of statutory power.

---

128   James E. Combs & Dan D. Nimmo, *The Comedy of Democracy* (Westport, CT: Praeger, 1996), 152.

8

A Legal Procedure to Protect Humankind

Destiny is not a matter of chance: it is a matter of choice. It is not a thing to be waited for; it is a thing to be achieved.

—William Jennings Bryan, "America's Mission," a speech delivered by the leader of the Democratic Party at the Washington Day banquet given by the Virginia Democratic Association at Washington, D.C. (February 22, 1899)

Is there a way to beat the system in which the triumvirate holds an upper hand on *freedom to* carry out their individual or group missions on development of superintelligence and radically enhanced humanoids? Why not consider appealing to a world level organization such as the United Nations to protect our human rights? Too high? Just exactly who would take on this task? Perhaps contact the organizations in the subsection titled International Policy Guidelines on Artificial Intelligence in Chapter 6? Can't there just be an international ban on some technological software? Specifically, AI tech companies? They have exorbitant power, don't they? How about an internationally exalted legal procedure?

The Debate on Regulations, Controls, and Bans

In 2016, U.S. tech companies had a concern about regulators jumping in to create rules around their AI work. At first, their efforts were focused on creating a framework for self-policing. Alphabet (Google's parent company), Amazon, Facebook,

IBM, and Microsoft, five of the world's largest tech companies, then began attempts to create an ethics standard around the creation of artificial intelligence. Their aims make clear that their research will benefit society, not harm it. Soon the group grew to more than 80 partners, including nonprofit research groups such as the Allen Institute for Artificial Intelligence.

By 2018, tech companies had admitted there were significant flaws in the self-policing system. Tech leaders then publicly acknowledged that their products might be flawed and harmful to employment, privacy, and human rights. Existential risk was, and still is, on how to reverse known and unknown outcomes. A Brookings report advised organizations to create a remediation plan for reversing or stopping environmental damage in case their AI technology does eventually inflict social harm. As the issue became clearer, calls went out for artificial intelligence ethics specialists.

Software giants Microsoft Corp. and Salesforce.com Inc. then hired ethicists to vet "unintended consequences that could result in a public relations fiasco or a legal headache."[129] So, what are the problems that have been facing the ethicists? For one thing, AI algorithms may be flawed. For another, Google had similar language for new products and ones that use AI and machine learning that raise new or existing ethical, technical, legal, and other challenges. This is just the beginning of a debate on regulations and bans on AI.

Authors of the Stanford report *Artificial Intelligence and Life in 2030* argue it will be impossible to regulate AI as a whole. "The study panel's consensus is that attempts to regulate AI in general would be misguided, since there is no clear definition of AI (it isn't one thing), and the risks and considerations are very different in different domains."[130] That report attempts to define the issues that citizens of a typical culture will face in robotic systems and computers that mimic human capabilities.

Matt Ridley, author of *The Evolution of Everything: How New Ideas Emerge*, argues that it is impossible to imagine all software development coming to a halt, even if the U.N. tried

---

129   John Murawski, "Need for AI Ethicists Becomes Clearer as Companies Admit Tech's Flaws," *WSJPRO*, March 1, 2019, https://www.wsj.com/articles/need-for-ai-ethicists-becomes-clearer-as-companies-admit-techs-flaws-11551436200.

130   John Markoff, *Seattle Times*, September 4, 2016, http://www.seattletimes.com/business/technology/real-ethics-for-artificial-intelligence/.

to enforce a ban on software development. The reason is twofold: technology is a sentient being, and the world is a large collection of disparate countries and scientists. Undeniably, there is no hope of stopping technological development per se globally in one big swoop. However, Ridley points out, "It is easier to prohibit technological development in larger-scale technologies that require big investments and national regulations. So, for example, Europe has fairly successfully maintained a de facto ban on genetic modification of crops for two decades in the name of the 'precautionary principle'—the idea that any possibility of harm, however remote, should scuttle new technology—and it looks as if it may do the same for shale gas."[131] Enter the precautionary principle that provides an opening for citizens, the government in the United States, and other countries to regulate the creation and development of detrimental technology that both has existential risk and raises issues about jobs lost to AI.

Creation and Legal Standing of the Precautionary Principle

The precautionary principle evolved out of the German socio-legal tradition, created in the heyday of democratic socialism in the 1930s; it centers on the concept of good household management. Through the late 1970s and early 1980s, the notions of care and wise practice evolved into focus on the environmental arena as the cornerstone for attention. "That focal point," wrote Sonja Boehmer Christiansen,[132] led to six basic concepts now enshrined in the principle:
- Preventative anticipation
- Safeguarding of ecological space
- Proportionality of response or cost-effectiveness of margins of error
- Duty of care, or onus of proof on those who propose change

---

131   Matt Ridley, "The Myth of Basic Science," *Wall Street Journal*, October 23, 2015, http://www.wsj.com/articles/the-myth-of-basic-science-1445613954.

132   As quoted in *Interpreting the Precautionary Principle*, Tim O'Riordan and James Cameron, eds. (London: Earthscan Publications Ltd., 1994).

- Proposing the cause of intrinsic natural rights
- Paying for past ecological debt

The primary concerns were for economic, moral and ecological considerations. The principle, as formalized, has expanded to include economic interpretation to ensure the protection of critical capital by deliberate management. The principle initially, and so far, has primarily been used on cases related to the physical environment. However, the viewpoint on its use has been changing for some time.

Tim O'Riordan and James Cameron had this to say in their seminal book *Interpreting the Precautionary Principle*, published in 1994: "This book will make clear that the precautionary principle is neither a single dimensional concept nor clearly defined even in its many features. It is entering the environmental arena in many ways, not all of which are obvious. Few of the openings contain traceable pathways suggesting where elements of precaution might lead. It is useful to contemplate the conditions through which precaution is likely to gain ground, and to envision circumstances where it might be resisted."[133] They end by writing, "This is a story of exploration in settings that are more favorable to change than many realize, but where the overall objective is still shrouded in mist. Precaution is here to stay, but its success may lie in its modesty and its preparedness to be transformed into various facets of social and political change without worrying about losing its name."[134]

After endorsement by the United Nations General Assembly, implementation of the precautionary principle was followed by the 1987 Montreal Protocol. The principle was then integrated with many other legally binding international treaties. Two examples are the (a) Rio Principles adopted by the 1992 Rio Conference Declaration on Environment and Development and (b) the Kyoto Protocol. The principle has been implemented in numerous agreements and declarations, and application of it has been made a statutory requirement in some countries[135] as well as in the European Union.

---

133 Ibid., 25.

134 Ibid., 28.

135 Art. 191 (2) TFEU, Explanations Relating to the Charter of Fundamental Rights (2007/C 303/02, OJ EU C303/35 14.12.2007 explanation on article 52 (5) of the EU Charter of Fundamental Rights, T-13/99 Pfizer vs Council, pp.114-125.

A significant number of other countries, including Australia, Japan, Brazil, Costa Rica, Argentina, Peru, and Ecuador have incorporated the precautionary principle as law.[136] America is a rare exception in that precautionary principle cases do not have the power of law.

The Commission of the European Communities noted in a February 2000 communication from the Commission on the Precautionary Principle that "the precautionary principle is not defined in the Treaties of the European Union, which prescribes it [the Precautionary Principle] only once—to protect the environment. But in practice, its scope is much wider, and specifically where preliminary-objective-scientific-evaluation indicates that there are reasonable grounds for concern that potentially dangerous effects on the environment, human, animal or [and] plant health may be inconsistent with the high level of protection [for what] chosen for the Community."[137] Note that there is no definition of what the term *health* entails. However, it would seem reasonable that existential risk to humankind would be included.

A difficulty is that although the principle does have strong legal standing through multiple treaties and nontreaty declarations,[138] application of the principle is somewhat hampered by a wide range of interpretations. In fact, "one study identified fourteen different formulations of the principle in treaties and nontreaty declarations."[139] In one of these, "R. B. Stewart (2002) reduced the precautionary principal to four basic versions."[140] They are:

1. Scientific uncertainty should not automatically preclude regulation of activities that pose a potential risk of significant harm (Non-Preclusion PP).

---

136   Christian Leathley and Elena Ponte, "The Evolution of the Precautionary Principle in Latin America," Legal Briefings, Herbert Smith Freehills, October 31, 2016.

137   Commission of the European Communities, Communication From The Commission on the Precautionary Principle, February 2, 2000.

138   Kenneth R. Foster, Paolo Vecchia, and Michael H. Repacholi, "Science and the Precautionary Principle," *Science* 288, no. 5468 (May 12 2000): 979–981, doi: 10.1126/science.288.5468.979.

139   Ibid.

140   R. B. Stewart, "Environmental Regulatory Decision Making Under Uncertainty," *Research in Law and Economics* 20 (2002): 76.

2. Regulatory controls should incorporate a margin of safety; activities should be limited below the level at which no adverse effect has been observed or predicted (Margin of Safety PP).
3. Activities that present an uncertain potential for significant harm should be subject to best technology available requirements to minimize the risk of harm unless the proponent of the activity shows that they present no appreciable risk of harm (BAT PP).
4. Activities that present an uncertain potential for significant harm should be prohibited unless the proponent of the activity shows that it presents no appreciable risk of harm (Prohibitory PP).[141]

There are also a number of other foundations for the legality of the principle. The 1998 Wingspread Statement on the Precautionary Principle summarizes it thus: "When an activity raises threats to human health or the environment, precautionary measures should be taken even if some cause and effect relationships are not fully established scientifically."[142] The January 2000 Cartagena Protocol on Biosafety, in regard to controversies over GMOs, stated: "Lack of scientific certainty due to insufficient relevant scientific information ... shall not prevent the Party of [I]mport, in order to avoid or minimize such potential adverse effects, from taking a decision, as appropriate, with regard to the import of the living modified organism in question."[143]

An example of how the precautionary principle works through the Cartagena Protocol can be seen in an international petition filed May 17, 2013, to stop the planting of genetically engineered Bt eggplants in Philippine test fields. The court "is the first in the world to adopt the precautionary principle regarding GMO products in its decision."[144] The rationale was

---

141 Wikipedia, s.v. "Precautionary principle," accessed January 2, 2016, https://en.wikipedia.org/wiki/Precautionary_principle.

142 Staff, Science and Environmental Health Network, Wingspread Conference on the Precautionary Principle, January 26, 1998.

143 "Cartagena Protocol on Biosafety to the Convention on Biological Diversity," United Nations, January 29, 2000, https://treaties.un.org/doc/Treaties/2000/01/20000129%2008-44%20PM/Ch_XXVII_08_ap.pdf.

144 Wikipedia, s.v. "Precautionary principle," accessed January 2, 2016, https://en.wikipedia.org/wiki/Precautionary_principle.

that the impacts of that undertaking on the environment, native crops, and human health are unknown. The Supreme Court, on December 8, 2015, permanently stopped the field testing. The ending of the Bt eggplant case is quite extraordinary, as the court reversed the decision as moot because the permits had expired.

THE PRECAUTIONARY PRINCIPLE IN THE UNITED STATES

By the turn of the century, the United States and the European Union had clashed over the regulation of a number of health and environmental risks, from genetically engineered food, to climate change, to beef.[145] In recent years there has been a movement in Europe to make the precautionary principle an overarching principle to govern all risk regulation. Some even assert that "the precautionary principle may already be so widely adopted that it is ripening into an enforceable norm of 'customary international law' from which no nation can dissent."[146] A backbone of European thinking about the precautionary principle is "prevention is better than cure."[147]

A larger and critical debate is about the proper stance of government: how should regulators act in the face of uncertainty about risk? "Today (2002), the conventional wisdom is that Europe endorses the precautionary principle and seeks proactively to regulate risks, while the US opposes the precautionary principle and waits more circumspectly for evidence of actual harm before regulating."[148]

There can be misunderstandings about appropriate use of

---

145 Jonathan B. Wiener, and Michael D. Rogers, "Comparing Precaution in the United States and Europe," *Journal of Risk Research* 5, no. 4 (2002): 317-349.

146 P. H. Sand, "The Precautionary Principle: A European Perspective," *Human and Ecological Risk Assessment* 6 (2000): 445-458.

147 Business Dictionary, http://www.businessdictionary.com/definition/precautionary-principle.html.

148 European Commission, "Second Annual Survey," "EU/Environment," *Agence Europe*, January 30, 1997, as quoted by Jonathan B. Weiner, "Convergence, Divergence, and Complexity," in *Green Giants? Environmental Policies of the United States and the European Union*, Norman J. Vig, Michael Gerbert Faure, Michael E. Kraft, and Sheldon Kamieniecki, eds. (Cambridge: MIT Press, 2004), 81.

the precautionary principle. One is that it is not meant for macro policy, which is concerned with monetary, economic growth; inflation; and national employment levels. That is clear from the reduced four versions of the precautionary principle by R. B. Stewart, which make plain that the use is for specific activity regarding rule of law and litigation. It is a guide, but not a macro guide.

Jonathan Adler, in the American Enterprise Institute paper "The Problem with Precaution: A Principle without Principle," argued, "It's better safe than sorry. We all accept this as a commonsense maxim. But can it guide public policy? Advocates of the precautionary principle think so, and argue that formalizing a more 'precautionary' approach to public health and environmental protection will better safeguard human well-being and the world around us. If it were only that easy." He continues, "Simply put, the precautionary principle is not a sound basis for public policy. At the broadest level of generality, the principle is unobjectionable, but it provides no meaningful guidance to pressing policy questions. In a public policy context, 'better safe than sorry' is a fairly vacuous instruction." Further, he notes, "Efforts to operationalize the precautionary principle into public law will do little to enhance the protection of public health and the environment." His conclusion is that "when selectively applied to politically disfavored technologies and conduct, the precautionary principle is a barrier to technological development and economic growth."[149] That is true, as stated by R. B. Stewart.

However, irreducible conflict between different interests is one of the difficulties with the precautionary principle, as debates in the United States do involve politics. The United States has its own problems concerning environmental risk assessment. Angela Logomasini, in her paper "EPA's Flawed IRIS Program Is Far from Gold Standard" argued that the United States Environmental Protection Agency's Integrated Risk Information System (IRIS) has a long history of sloppy research. Further, it has evidenced a "lack of transparency that

---

149 Jonathan Adler, "The Problems with Precaution: A Principle without Principle," American Enterprise Institute (AEI), May 25, 2011, https://www.aei.org/articles/the-problems-with-precaution-a-principle-without-principle/.

has advanced faulty and often counterproductive regulations that impose needless burdens on the public."[150]

An example of a political conflict problem can be found in an endangerment finding whether the Environmental Protection Agency (EPA) has legal authority under the Clear Air Act to regulate carbon dioxide emissions—in particular, whether the EPA has discretion *not* to regulate carbon dioxide emission for policy reasons under the Clean Air Act. That was released in the EPA's 2009 legal opinion, which concluded that, because of its contribution to global warming, carbon dioxide in large amounts "met the Clean Air Act's definition of a pollutant that harms human health."[151] It is the events leading up to final admittance that the EPA has authority to regulate the dioxide emissions that is the interesting and useful part.

The text of the Clean Air Act was first passed in 1963 before being amended in 1970, 1977, and 1990, and it does not refer to climate change or greenhouse gases. In 1999, Massachusetts and 11 other states petitioned the EPA for regulation of carbon dioxide (CO_2) as an air pollutant. In 2003, the EPA denied the petition, arguing it did not have the legal authority under the Act to regulate carbon dioxide and other, similar gases as air pollutants. Now the political conflict: "In addition, the EPA argued that if it did have the authority it would not regulate carbon dioxide because it would interfere with the George W. Bush administration's preferred policy approach to human-caused climate change."[152]

"Massachusetts appealed to the United States Court of Appeals for the District of Columbia Circuit in 2005, which upheld the EPA's position. Massachusetts then appealed to the Supreme Court of the United States, which heard the case in November 2006. The court issued its ruling in April

---

150   Angela Logomasini, "EPA's Flawed Iris Program Is Far from Gold Standard," Competitive Enterprise Institute, February 12, 2019, https://cei.org/content/epas-flawed-iris-program-far-gold-standard.

151   The EPA made its formal finding that greenhouse gases endanger the public health and welfare in 74 Federal Register 66496 (Dec. 15, 2009).

152   Ballotpedia, "Massachusetts v. Environmental Protection Agency," accessed November 4, 2020, https://ballotpedia.org/Massachusetts_v._Environmental_Protection_Agency.

2007 and sided with" Massachusetts.[153] This is an example of a case in which the precautionary principle might well be used. However, the legal side of the principle is often hampered by both lack of political will and/or extreme power of, and within, the government, because the principle does not have statutory power in the United States. The most important point in this example about the EPA travails is that legal cases can drag on and on almost indefinitely. A critical point for this book is the need to allow a seemingly extraordinary amount of time in preparation and implementation of precautionary principle court cases—in particular, cases with existential risk, as is the situation in ones regarding superintelligence and extensively augmented humanoids.

Use of The Precautionary Principle in Research and Development Cases

Existential risk from superintelligence development will inevitably manifest itself—quickly in many cases—without sufficient warning for people to take remedial action. As of 2017, over forty organizations worldwide are doing active research on artificial general intelligence (AGI). That refers to the ability to accomplish any cognitive task at least as well as humans, or to "a machine that has the capacity to understand or learn any intellectual task that a human being can."[154] Unfortunately, there are less responsible individuals and miscreants who can attempt to hack into the developers' work and algorithms to instill their own beliefs. The interest of some, notably singularitarians, is the creation of posthumanity conditions.

In short, considering the above situation, it would seem to be immoral to not give high priority to putting the precautionary principle into force on AGI research in the near term. However,

---

153 Supreme Court of the United States, "Massachusetts et al. c. Environmental Protection Agency et al.," October term 2006.

154 Seth Baum, "A Survey of Artificial General Intelligence Projects for Ethics, Risk and Policy," Global Catastrophic Risk Institute Working Paper 17-1, November 12, 2017, available at SSRN: https://ssrn.com/abstract=3070741 or http://dx.doi.org/10.2139/ssrn.3070741.

there are two caveats. First, a precautionary principle court case on the topic would have to be created. Second, a working committee would have to be established within a reasonable time frame in a location with statuary power where minimal political interference can be assured.

There are compelling arguments in the Asilomar Principles that risks posed by AI systems, especially catastrophic or existential risks, must be subject to planning and mitigation efforts commensurate with their expected impact. The precautionary principles do include a method to carry out that practice. Although it is not made explicit in all legislation, the EU Commission on the Precautionary Principle states that measures based on the precautionary principle should be periodically reviewed. They should then be amended as necessary when new evidence is available.

Some EU legislation (e.g., Directive 2011/65/EU40) expresses the provisional nature of precautionary measures arising from the precautionary principle and lays down a requirement for review in the light of new evidence, or a requirement to develop evidence. This requirement allows for assessing whether precautionary action has produced the intended consequences. It is also a method to check "whether measures put in place need to be modified, taking into account new information or knowledge that may reduce the degree of scientific uncertainty."[155]

Kenisha Garnett and David Parsons reviewed fifteen cases in a study on EU and member states. They found there was a lack of guidance on what conditions justify a reexamination of the potential risks and who would be responsible for producing the evidence required for risk assessment. The conclusion was that, despite guidance not being perfect, there is at least an international system in which there are legal procedures for risk assessment and control.[156]

The United States is the logical site for a working group, considering the extent of research and development on AI in that country. But there are drawbacks. At the very least, the working committee would have to have statutory authority and

---

155 Kenisha Garnett and David J. Parsons, "Multi-Case Review of the Application of the Precautionary Principle in European Union Law and Case Law," *Risk Analysis* 37, no. 3 (March 2017): 502-516.

156 Ibid.

wide recognition of the precautionary principle, conditions that the United States is not willing to grant at this time. Considering that legal proceedings will be a lengthy process, if the United States does not give high priority to that requirement, other venues will have to be selected.

Many technologies carry numerous risks that range from economic to moral. For instance, the race to develop superintelligence is currently a subject of much scientific uncertainty. A similar state of affairs applies to humanoid developers who have no endgame in sight. At this point, science does not have sufficient information to rule on the likelihood or even the theoretical possibilities of such risks. Considering that, the precautionary principle, which implies there is social responsibility to protect the public from exposure to harm when scientific investigation has found plausible risk, rests on two core ideas:

> (1) An expression of a need by decision-makers to anticipate harm before it occurs. Within this element lies an implicit reversal of the onus of proof. Under the precautionary principles the proposed activity will not (or is very unlikely to) result in significant harm.
> (2) The concept of proportionality of the risk and the cost and feasibility of a proposed action."[157]

Succinctly, this concept of the principle states that if a *proposed action or policy*, i.e. *one that has not yet begun*, has a suspected risk of causing harm to the public or to the environment, in absence of scientific consensus that the action or policy is not harmful, the burden of proof that it is *not* harmful falls on those wanting to proceed with the activity or project. A key basic concept in the principle is that "protections can be relaxed only if *further* scientific findings provide sound evidence that no harm will result."[158]

---

157   Andrew Jordan and Timothy O'Riordan, "The Precautionary Principle: A Legal and Policy History," in *The Precautionary Principle: Protecting Public Health, the Environment and the Future of Our Children*, Marco Martuzzi and Joel Ticker, eds. (Copenhagen: The World Health Organization, 2004), http://www.euro.who.int/en/publications. Go to search, 0003/91173/E83079.pdf.

158   Paul Sloane, "How the Innovation Principle Supplements and Advances the Precautionary Principle," Innovation Management, https://innovationmanagement.se/2017/05/31/how-the-innovation-principle-supplements-and-balances-the-precautionary-principle.

Almost any action creates a certain amount of risk, and this is especially true of research that seeks to answer unsolved questions. There are two generally accepted forms of the precautionary principle, defined as *strong* and *weak* in the international agreement and declarations on scientific uncertainty. The *strong* option holds that *regulation is required* whenever there is possible risk to health, safety, or the environment even if the supporting evidence is speculative, and even if the economic costs of regulation are high.[159]

The *weak* form in the international agreement holds that lack of scientific evidence on the pros and cons of research *does not preclude (stop or rule out)* action to let the research *continue*. Under weak formulations, *the requirement to justify the need for action*, i.e. to impose regulations, controls or bans on activities (the burden of proof) *generally falls on those advocating precautionary action*.

Inaction is a term used for determination that the activity should *not proceed*, the reason being that inaction carries its own risks, which may be greater than the risks of action. *Action* is a term for determination that the activity be allowed *to proceed*.

Extension of the precautionary principle to research, technological creation, and development is basically new territory, for three reasons. One is that, contrary to much of historical use of the principle, projects like superintelligence and humanoid development are *already in motion*. A second is that in *ongoing research and development*, uncertainty, and existential risk to humankind play a major role. A third is the use of cost-benefit analysis on very large projects.

Precautionary Principle Use in Artificial Intelligence Projects

A critical question: How can AI be corralled? Some believe it shouldn't—that we should be moving full steam ahead. Others, and I am one of them, are convinced there should be regulations, controls, or bans given the existential risk from superintelligence. Why? For moral and economic

---

[159] Italics in this section are the author's for emphasis.

considerations. Yet a wide scope of individuals are convinced they have *freedom to* create any and all sorts of innovations. On the *freedom from* side is the issue of global job losses arising from unconscionable developments in artificial intelligence.

Think back to Matt Ridley, who argues that although there is no hope of stopping all technology development per se, "It is easier to prohibit technological development in larger-scale technologies that require big investments and national regulations." Recall that the precautionary principle is not appropriate for macro-level policy projects. It is not appropriate for bundling artificial intelligence together as a whole. Nor is appropriate for coupling several diverse AI segments. Nevertheless, the precautionary principle is an appropriate modus operandi to adjudicate on what and the way in which general AI and robotics are changing people's lives. The idea and objective in this section is to explain how cost-benefit analysis is used in cases like job losses.

There are several options if a group decides to initiate a case regulating job losses resulting from AI. A case could cover a small area and essentially be a test case as a preview to a larger project. The approach would be to begin within a short period of time, focusing on the protocol for a larger project. The concept could be to eventually have a project large enough to attain a decision in court that probably could have several parts to it. The methods and analysis protocol are for a large project that includes the impact of robotics; robots, humanoid robots, and true humanoids. This is also the kind of situation the precautionary principle can be used for, to address the combination of threat of harm and uncertainty. This section does not go into details. Rather, the objective is to conceptualize the use of one major tool.

COST–BENEFIT ANALYSIS

Cost–benefit analysis (CBA) is often used by organizations to appraise the desirability of physical projects and given policies. It is an analysis of the expected balance of costs and benefits, including an account of any alternatives and the status quo. CBA helps predict whether the benefits of a policy or nonpolicy projects outweigh its costs (and by how much), relative to other alternatives. This allows the ranking of alternative policies in terms of a cost–benefit ratio.

Generally, accurate cost–benefit analysis identifies choices which increase welfare from a utilitarian (practical) perspective. Although CBA can offer an informed estimate of the best alternative, a perfect appraisal of all present and future costs and benefits is difficult; perfection, in economic efficiency and social welfare, is not guaranteed.[160]

That issue is a principal factor discussed in this example that focuses on worldwide job losses.

CBA has been used in precautionary principle cases for a long time, although the one given here is by far the most complex. Separation of costs and benefits between robotics and general AI is essential. Job losses in each area diagnosed complicates this cost–benefit analysis. Variables and parameters keep changing in short times. There are cultural and ethnic considerations across borders regarding how to treat the impact. One most difficult and time-consuming task is calculating lost income across multiple geographic areas. And then the extent to which workers could return to jobs, and ones permanently lost. Also, compensation for opportunity costs, if that becomes an issue along with others. There is uncertainty and risk of projections. However, this kind of multiplicity in conditions (at a much lower level) is what the international aspect of the precautionary principle is meant to cover.

In all likelihood, considering the rapidity of AI development, governmental projections on the economy, and dire forecasts of job losses resulting from AI around 2030, some form of the *strong* option would likely be in order from results of the CBA. In simple terms, it means that regulations, controls, and bans would be placed in such a way that there would be continuance allowed on technologies that reduce job losses, but would include some positive benefits, such as new types of positions for workers.[161] Given only this cursory explanation of cost–benefit analysis and the problem, would it seem reasonable to think the cost–benefit ratio would be positive?

---

160   Wikipedia, s.v. "Cost–Benefit Analysis," accessed September 12, 2020, https://en.wikipedia.org/wiki/Cost-benefit_analysis.

161   Cass R. Sunstein, "The Paralyzing Principal: Does the Precautionary Principle Point Us in Any Helpful Direction?" Winter, 2002-2003, The Cato Institute.

As we move to the next chapter and the role of human rights, let's reflect on the topic of human rights advocates and other equally concerned individuals and groups on animal rights. That is a subject in Chapter 3. Recall that, given that there are already fifty or so animal rights organizations, there is likelihood of lobbying to provide rights similar to those of humans for these creatures as well as for humanoids. This is a seemingly weird situation where requests for precautionary principle court hearings could test what Tim O'Riordan and James Cameron meant when they wrote, "Precaution is here to stay, but its success may lie in its modesty and its preparedness to be transformed into various facets of social and political change without worrying about losing its name."[162]

---

162   *Interpreting the Precautionary Principle*, Tim O'Riordan and James Cameron, eds. (London: Earthscan Publications Ltd., 1994).

9

"Sorry, Your Job is Lost to Artificial Intelligence"

We live in a society exquisitely dependent on science and technology, in which hardly anyone knows anything about science and technology. This is a clear prescription for disaster.

—Carl Sagan
The Skeptical Inquirer (Volume 14-3, Spring 1990).

The news is not so good. The novel SARS-CoV-2 coronavirus, which resulted in the large scale COVID-19 pandemic, has overpowered the already tough times for the industrial sector of our economy. A large-scale recession now looms. One of the natural phenomena of recessions is creative destruction, which the great economist Joseph Schumpeter popularized as an essential fact about capitalism. He postulated that the structural adjustment following recessions leads to growth in jobs, increases in wages, and elevated standards of living. Those readjustments have indeed taken place, but they are relics of past times. The Great Recession that began in December 2007 and ended in June 2009 has been different from all previous ones. Wages and standards of living did not improve. The forecast regarding the recession anticipated in 2021 is bad. Regrettably, exponential technology developments and other economic forces have tragically restricted net improvements in living standards. Can anything else happen to make matters worse?

Yes, the Great Race to Technologize America Does Result in Job Losses

Between 2000 and 2010, nearly 6 million American factory workers lost their jobs. Manufacturing employment plummeted by a third. By 2018, employment was still 26 percent smaller than in 2000, with only about 13 million jobs left. Part of the blame falls on growing trade deficits as low-wage nations can produce products at lower cost. However, AI-based technology adoption is a predominant factor in job losses. There are arguments from several sides that goods-producing jobs will grow and help make America great again. However, the reality is that the number of people employed in the goods-producing side of the industry sector, which includes manufacturing, is now relatively small. The non-farm workforce dropped from 40 percent in 1960 to around 17 percent in 2016 (Graph 1).

Manufacturing, which constituted 28 percent of the United States workforce in 1960, accounted for just 9 percent in 2017. As if that is not bad enough, the gig economy is increasingly prevailing, in which large numbers of people work part-time or temporary positions. In this structure, a significant portion of the goods-producing employees are low-wage, insecure, on-demand temps. They, as well as part-time jobs, have few, if any, benefits. The "hope all goes well" story on America's manufacturing revival rests on three flimsy exigencies:

- that consumers awaken to the notion that making goods in America means jobs for Americans;
- that companies come to understand that home-grown intellectual property must be protected; and
- that production will grow faster than the general economy.

Three pivotal elements are needed to drive a revival of American manufacturing: solid profits, rising productivity, and wage growth. Of the three, American manufacturing only has solid profits, which are insecure. Bottom line: hope is not enough. We have to face facts, regardless of whether we like the scenarios.

Graph 1: Goods-Producing and Service Sector Jobs, United States 1940–2018

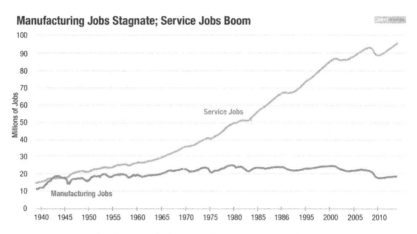

Source: Bureau of Labor Statistics. Credit: Quoctrung Bui/NPR . https://www.npr.org/sections/money/2014/01/02/259131399/where-the-jobs-are-in-2-graphs-hint-not-in-manufacturing

Every country's economy is composed of three sectors: industry, services, and agriculture. The most common collective measurement of the three sectors is gross domestic product (GDP), the monetary value of all the finished (final) foods and services produced within a country's borders in a period of time, usually a year. GDP is important because it gives information about the size of the economy and how it is performing. The growth rate of real GDP is often used as an indicator of the general health of the economy. The increase in real GDP in broad terms is interpreted as a sign that the economy is doing well. Industry, of which manufacturing is a part, has large economic multipliers (measures of direct as well as indirect effect in an economic activity) that faced a 24 percent decline in GDP from 2010 to 2018.

Sadly, the forecast for manufacturing and the economy as a whole for the next decade is not auspicious. The United States had a debt-to-GDP ratio of 105.4 percent in 2017. By 2020, the ratio was 106.2 percent. To put these figures into perspective, the U.S.'s highest debt-to-GDP ratio was 121.7 percent at the end of World War II, in 1946. In comparison, among the developed countries, the U.S.'s ratio is third highest. Italy's debt in 2020 is 133.2 percent and Japan's is 237.6 percent.

The Congressional Budget Office (CBO), in the August 2019 annual outlook for the next ten years, estimated the national debt would surge from 78 percent of GDP to 92 percent, driving the debt-to-GDP to an unthinkable level. That was before the novel COVID-19 epidemic struck. By March of 2020, the public national debt had soared from around $22 trillion to around $23.7 trillion. The interest alone in 2019 was $404 billion and is rising at a brisk pace and has become one of the most prominent political issues.

Pundits Project Massive Job Losses in the Longer Term

Longer term projections by experts on job losses are equally discouraging. The consulting firm Pricewaterhouse Cooper (PwC) estimates 38 percent of U.S. jobs could be automated by 2030.[163] However, if it helps much, the firm found that in many cases people won't lose their jobs outright—they will just be shuffled around. There are 125 million employees in the U.S. As incredible as it may seem, if the PwC's estimate of 38 percent of U.S. jobs being automated by 2030 is anywhere near correct, that means roughly the 47 million people out of work from their initial jobs will struggle to find other sources of employment. PwC is predicting the loss of 30 percent of jobs in the U.K. Gartner Inc., the technology research firm, has predicted that one-third of all jobs will be lost to automation within a decade.

The worry is not just in the United States and Europe. The International Labour Organization (ILO) in its 2016 report "ASEAN in Transformation: The Future of Jobs at Risk of Automation" found that nearly three in five jobs in that region face a high risk of automation in the next couple of decades.[164]

---

163   Samatha Masunaga, "Automation could replace one out of three US jobs within about 15 years, report says," *Seattle Times*, March 27, 2017, https://www.seattletimes.com/business/technology/automation-could-replace-one-out-of-three-us-jobs-within-about-15-years-report-says/.

164   Jae-Hee Chang and Phu Huynh, *ASEAN in Transformation: The Future of Jobs at Risk for Automation*, International Labour Office, Bureau for Employers' Activities; ILO Regional Office for Asia and the Pacific, Geneva, ILO Working Paper No. 9, July 2016, available at https://www.ilo.org/wcmsp5/groups/public/---ed_dialogue/---act_emp/documents/publication/wcms_579554.pdf.

The McKinsey Global Institute reported that current technologies can lead to automation of almost half of paid workers jobs by the 2030s. Economists at Oxford University forecast that machine technology will account for nearly half of current jobs within two decades.[165]

Another of McKinsey & Co.'s predictions is that investment in technology, including AI and automation, could contribute to the loss of 20 million to 50 million jobs *globally* [emphasis mine] by 2030. That means that not only would a huge percent of jobs be lost, but an enormous number of people from all sectors would need to switch occupations. That is provided jobs even become available. Virtually all of the drop out in the workforce by 2030 will result from the adoption of automation.[166]

Here is a critical thinking matter. Just exactly what kind of jobs, if any, will be available?. Some pundits opine that AI opens up other opportunities, such as the need for new human workers to keep the smart world running smoothly by managing and collaborating with AI in some form. Some believe that in the long term, jobs will increase in high-tech companies, or in ones derived from AI innovations and other creations. But that seems unlikely.

Consulting firm McKinsey & Co. reported the number of new technology jobs it expects will be created in the U.S. by 2030 through investment in AI, automation, and related technologies as follows:[167]

- Software application developers – 289,000
- System software developers – 111,000
- Computer system analysts – 108,000
- Computer programmers – 84,000
- Total – 686, 000

Distressingly, 686,000 is a trivial amount compared to the 47 million jobs lost under PwC's projections.

---

165 Timothy Aeppel, "What Clever Robots Mean for Jobs," *Wall Street Journal*, February 24, 2015, http://www.wsj.com/articles/what-clever-robots-mean-for-jobs-1424835002015.

166 Ibid.

167 Daniela Hernandez, "Seven Jobs Robots Will Create—Or Expand," *Wall Street Journal*, April 30, 2018, https://www.wsj.com/articles/seven-jobs-robots-will-createor-expand-1525054021.

An April 2018 *Wall Street Journal* article titled "Seven Jobs Robots Will Create—or Expand" projects that the following jobs will be created:[168]

1. AI builders that develop the underlying systems that make AI work
2. Customer-robot-liaisons such as ensuring that clients are happy with robot security guards
3. Robot managers that oversee the work of machines such as robots (and I add humanoid robots)
4. Data labelers for looking over information and marking it for a computer
5. Drone-performance artists that customize drones to perform different performances
6. AI lab scientists that teach AI machines or algorithms about life sciences or chemistry so that computers can surface novel ideas
7. Safety and test drivers for vehicles

The takeaway from the two lists is that added jobs are trifling, considering the projected 47 million jobs lost to automation in America alone. Experts as panelists at an April 2018 MIT Digital Economy Conference[169] explained the simple reason why significantly more jobs won't be created by businesses: companies have yet to embrace AI. To the panelists, the problem is that companies are just not up to speed with *machine learning*, an "application of AI that provides systems with the ability to automatically learn and improve from experience without being explicitly programmed."[170]

Currently, there is a dearth of properly skilled employees and hires to teach robots and humanoid robots how to develop new skill sets to fill the gaps.[171] In March 2019, the consulting

---

168 Ibid.

169 Steven Norton, "Machine Learning at Scale Remains Elusive for Many Firms," *Wall Street Journal*, April 27, 2018, https://blogs.wsj.com/cio/2018/04/27/machine-learning-at-scale-remains-elusive-for-many-firms/.

170 Expert System Team, "What is Machine Learning? A definition," blog post, May 6, 2020, https://expertsystem.com/machine-learning-definition/#:~:text=Machine%20learning%20is%20an%20application,use%20it%20learn%20for%20themselves.

171 John Murawski, "Businesses' Adoption of AI Is Expected to Surge," *WSJPRO*, April 25, 2019, https://www.wsj.com/articles/businesses-adoption-of-ai-is-expected-to-surge-11556184602?mod=djemAIPro.

firm Deloitte LLP said that within two years, 72 percent of businesses will have implemented or be planning to implement the technologies to gain insights in core functions. To me, the objective is really to cut costs, which in turn tragically translates into dismissing employees.

There are occasional stories in the media suggesting that, based on history, technological AI will benefit today's workers through automation. One is that of Dr. Atkinson of the Information and Innovation Foundation. He maintained in April 2019 that "AI will benefit today's workers who, if they lose a job to automation, can more easily find another as AI fuels more growth."[172] But then came a non sequitur: "To be sure, AI is much less likely to complement—and more likely to replace—less skilled workers. But this should be seen as good news, since the U.S. economy will have relatively fewer low-wage jobs and relatively more middle-and higher-wage jobs. None of this is to say the U.S. shouldn't do more to prepare workers, especially lower-skilled ones, for transitions into new jobs and occupations. But those are manageable problems that policy makers can solve, rather than succumbing to a panic about a threat that isn't real."[173]

Job Losses by White-Collar and Agricultural Workers

Less recognized, thanks to a historically low unemployment rate and a historically long bull market, are job losses in America's service sector. These range from tellers at banks to toll takers, primarily because the business community worships the god of efficiency. Robots are now becoming a deciding factor in who gets jobs and whom to fire for underperformance. Amazon tracks the productivity of its employees at fulfillment centers with its Associate Development and Performance Tracker,

---

172   Dr. Atkinson, "It Will Lead to More Spending and Investing—and Jobs" in a debate with Dr. Frey, who argued, "History Tells Us the Answer, and It Isn't Encouraging," in "Will AI Destroy More Jobs than It Creates over the Next Decade," *Wall Street Journal*, April 1, 2019, https://www.wsj.com/articles/will-ai-destroy-more-jobs-than-it-creates-over-the-next-decade-11554156299.

173   Ibid.

termed Adapt. Managers make the final decisions; yet, as Ian Larkin, a business professor at the University of California at Los Angeles, said, the discipline process "makes an already difficult job seem even more inhuman and undesirable." [174]

Technology has also taken, and continues to take, the place of auto workers, file clerks, ticket agents, state and national regulators, academics, and even the legal profession. Some authorities maintain that within a decade or two, even CEOs will begin to find their jobs outsourced. Losing one's job is bad enough; discovering that one's resume was not selected by a bot taught to use resume-filtering software has to be much more than frustrating. The culprit, as job losses accelerate while the American economy reverses course from the boom years, is, in addition to worship of the god of efficiency, worship of the gods of productivity and of cost effectiveness.

Robotics' moving from industry into agriculture, the third sector of economies, is becoming mainstream in our increasingly droid-oriented new world. For example, even in the agriculture sector, AI innovations and computers dominate our ever-larger farms. Robots are taking over at race pace. The first fully autonomous farm equipment is becoming commercially available in a way that analysts had not predicted. The ideal way to go, thanks to startups in Canada and Australia, is with small tractors that can perform thousands of tasks, while large tractors and equipment are limited. The idea is that to move up to the next level of profitability and efficiency, farmers downsize to autonomous tractors that prioritize precision in all tasks like seeding rates, water requirements, and chemicals. Add to that essentially no need for drivers and other employees, and the savings are significant.

Robots are also increasingly harvesting fruits and vegetables such as strawberries in which high-powered computing, color sensors, and small metal baskets attached to robotic arms gently pluck ripe berries. In short, goodbye field hand! So who are the big gainers in the fruits and vegetable areas? –the wealthy Silicon Valley tycoons and entrepreneurs that are buying up farmland, developing startups, and selling robots to work the

---

174 Greg Ip, "For Lower-paid Workers, the Robot Overlords have Arrived," *Wall Street Journal*, May 1, 2019, https://www.wsj.com/articles/for-lower-paid-workers-the-robot-overlords-have-arrived-11556719323

fields. Other types of businesses are also pushing their way into every stage of food-growing and -processing industries. Efforts are underway up the food supply chain to do away with human labor used for deboning chickens by replacing them with surgical robots in meat packing plants. The same holds for beef and pork; the coronavirus swept through packing plants because of tight working conditions. Robots are also being "employed" on dairy farms to milk cows and feed them. The results are improved yields, lower costs, and adios to human workers.

Now for the genuinely weighty downside. Rising efficiency in industry and commerce threatens to increase inequality in society. Edward Tenner, noting that cheaper and better machines and electronic platforms are coming, invokes what he terms the Efficiency Paradox, which supposes that "average people will benefit somewhat; the top will benefit tremendously."[175] The reviewer Gregg Easterbrook asks, "Does this mean rising efficiency should be banned? Mr. Tenner doesn't wager an opinion."

---

175   Gregg Easterbrook, "The Efficiency Paradox Review: Big Data, Big Problems," April 22, 2018, https://www.wsj.com/articles/the-efficiency-paradox-review-big-da-ta-big-problems-1524420853#:~:text=GreggEasterbrookreviewsE9CTheEfficiency-ParadoxEDbyEdwardTenner.&text='BigDataistheBig,thembetterthanwedo.

10

THE FUTURE OF JOBS

Life is not a search for experience, but for ourselves. Having discovered our fundamental level we realize that it conforms to our destiny and we find peace.

—Cesare Pavese, *This Business of Living*

The service sector has taken a back seat in discussions about the economy. Surprising is the wide number of economic and social service components. These include transport, storage, communication, trade, hotels, tourism, banking, insurance services, education, health, and administration. What is little known is how that sector has steadily grown to the extent that it accounted for 102.3 million jobs in 2016 (Graph 1). Startlingly, it accounted for 80.3 percent of employment in 2016 compared with 12.6 percent in the goods-producing sector. Even more startling are the significant social drawbacks in the service sector.

ENTER THE SERVICE SECTOR OF THE ECONOMY

Jobs that primarily lifted women are now dwindling, and they are not coming back. In fact, "the United States has shed more than 2.1 million administrative and office support jobs since 2000," according to Labor Department data. A pathetic element is that the Labor Department has also predicted that "secretaries and administrative assistants will see the largest job losses of any occupation in the coming decade." Most of the "administrative positions have been lost to outsourcing," whereas "'The new trend is to actually replace administrative work with robotic software,' said Phil Fersht, head of HFS Research." The grim part as of 2019 is that "among the 10 jobs expected to add the most employees in the next decade, six pay less than $27,000 a year."[176] The changes are especially hard on women. More than half of workers aged fifty and older are let go involuntarily. Only one in ten of those workers is expected to earn as much again.

There is yet another major hitch on the economic well-being side. Addition of many service sector jobs alone will not lead to long-term sustainable economic growth because service sector jobs have low economic multipliers.[177] That is because the sector is composed primarily of white-collar workers that range from teachers to hi-tech engineers, accountants, and computer specialists. They produce ideas, knowledge, and personal services, but largely do not produce physical things. Neither do services such as restaurants, movie theaters, Facebook, and overseas investments. Consequently, all of these have a relatively low impact on economic growth in the economy.

One example is furniture manufacturing in the city of Dallas, Texas, which has an economic multiplier of 2.73, while restaurants there have a multiplier of 1.5.[178] Coffeeshops do make a physical product, but the ingredients are minimal, resulting in a relatively low multiplier effect. In addition, the jobs are low income. In brief, industry, with its high jobs multiplier of economic benefits, has drastically declined, while the services sector, with smaller multipliers, has grown dramatically. Another aspect of the drag on the economy in the last few decades is slow growth in worker productivity—the amount of product turned out by a worker per unit of time.

---

176 Heather Long, "Administrative assistant jobs helped propel many women into the middle class. Now they're disappearing," *Washington Post*, December 5, 2019, https://www.washingtonpost.com/business/economy/administrative-assistant-jobs-helped-propel-many-women-into-the-middle-class-now-theyre-disappearing/2019/12/04/75686efe-f6a0-11e9-a285-882a8e386a96_story.html.

177 James Bessen, *Learning by Doing* (New Haven: Yale University Press, 2015).

178 Impact Data Source, "What's My Multiplier?" March 30, 2016, https://impactdatasource.com/whats-my-multiplier/.

The result is reduced payments and reductions in the national debt. Paradoxically, a touted solution for economic growth in highly developed, densely populated, and homogeneous Asian countries as well as in some European ones is increased use of robots. To an extent, that is free labor. Add to that the benefits of job diversification.

Let's shift to the economic side of the U.S economy. The news is awful. Early in 2019, the U.S. Congressional Budget Offices' (CBO) forecast was that economic growth would be a respectable 2.3 percent in 2019. That kind of level is gone. The January 2020 CBO's 2020 projection was that growth will average only 1.7 percent in the entire 2020–2030 decade. Even that level is provided there are no dramatic interruptions. There are other reasons to fear the future. The U.S. federal government debt is set to skyrocket over the next ten years from 16 trillion dollars in 2019 to 33 trillion dollars in 2030. A recession of devastating proportions is in the making. An impending severe calamity is that Social Security will run out of money in 2035 unless the government steps in. Medicare's hospital insurance trust will run out in 2026. That's just the beginning.

THE NOVEL CORONAVIRUS FACTOR

The First World War began on September 6, 1914, with the battle of the Marne, which the British and French won against the German troops advancing on Paris. That war ended in November 1918. The Spanish flu, also termed the H1N1 virus pandemic, officially began during that war in January 1918, and ended in December 1920. Dave Roos notes that "while the global pandemic lasted for two years, the vast majority of deaths were packed into three especially cruel months in the fall of 1918. Historians now believe that the fatal severity of the Spanish flu's 'second wave' was caused by a mutated virus spread by wartime troop movements."[179] The vulnerability of those healthy young

---

179   Dave Roos, "Why the Second Wave of the 1918 Spanish Flu Was So Deadly," March 3, 2020, updated April 29, 2020, https://www.history.com/news/spanish-flu-second-wave-resurgence.

adults, and lack of vaccines and treatments, created a major public health crisis. That pandemic wound up causing at least 50 million deaths worldwide, including approximately 675,000 in the United States. In contrast, and not well known, is that epidemics and pandemics are estimated to have killed 300–500 million people over the course of human history.

The first known severe illness caused by a coronavirus was the 2003 Severe Acute Respiratory Syndrome (SARS) epidemic, which emerged in China. A second outbreak of severe illness began in 2012 in Saudi Arabia with the Middle East Respiratory Syndrome (MERS). The novel SARS-CoV-2 coronavirus, which resulted in the large-scale COVID-19 epidemic that spread to more than seventy other countries, was found in the city of Wuhan, China, at the end of December 2019.

Thomas A. Garrett, assistant vice president and economist at the Federal Reserve Bank of St. Louis, released a report in 2007 titled *Economic Effects of the 2018 Influenza Pandemic: Implications for a Modern-day Pandemic*. He was prescient when he wrote at the beginning that "the possibility of a worldwide influenza pandemic in the near future is of growing concern for many countries around the globe." His grim conclusion: "Given our highly mobile and connected society, any future influenza pandemic is likely to be more severe in its reach, and perhaps in its virulence, than the 1918 influenza despite improvements in health care over the past 90 years." Most important, Garrett argued that it is the government's role and duty to have the readiness and ability to protect citizens from a pandemic, and that mitigating a pandemic requires cooperation and planning at all levels of government and the private sector.

He also observes, "Many predictions of the economic and social costs of a modern-day influenza pandemic are based on the effects of the influenza pandemic of 1918." He used that information to suggest that globally there would be an initial cost of several hundred billion dollars and the death of thousands to several million people. He then quotes Alfred W. Crosby's 2003 book, *America's Forgotten Pandemic: The Influence of 1918*,[180] which says that although the influenza

---

180   2013 Cambridge University Press, p. 22 in Garrett.

was short-lived, it "'had a permanent influence not only on the collectivities but on the atoms of human society—individuals.'"

Brian Doherty presented a study in *Reason.com*[181] from the 2013 *Journal of Health Economics* about Sweden, a country with a mostly homogenous population. The judgement from the detailed data on the 1918–20 pandemic is that 37,573 people died. Despite a significant increase in poverty rates and an economy that shrank dramatically, the recovery was very quick. GDP had increased by 8 percent in 1922, and there was continued steady real economic growth and wages for the rest of the decade. The Garrett report is in consensus with Crosby and Doherty that a recession was avoided.

The story has changed dramatically during the current pandemic in Sweden. The reason: trust in government, institutions, and fellow Swedes. An unorthodox, open-air experiment was conducted that did not impose lockdowns and allowed restaurants, gyms, shops, playgrounds and most schools to remain open. More than three months later, in early July, the coronavirus is blamed for more than 5,420 deaths, most of which could have been avoided. The Swedish economy is expected to contract by 4.5 percent this year.[182]

That brings us to the impact of the 2020 COVID-19 epidemic in the United States, which had resulted in 147,000 deaths by mid-July. On December 31, 2019, the government in Wuhan, China, confirmed that health authorities were treating dozens of cases. Days later, researchers in China identified a new virus that had infected dozens of people in Asia. At the time, there was no evidence that the virus was readily spread by humans. The greatest tragedy is there had been ample warning in the United States through continued strident requests from the national health community over the years to prepare for extreme flu and corona viruses. Nonetheless, our shortsighted, dysfunctional administration and president took on the task of playing catch-up only grudgingly, generally leaving the states to fend for themselves.

---

181   Brian Doherty, Reason.com, March 20, 2020, https://reason.com/2020/03/20/what-economic-analyses-of-past-pandemics-can-tell-us-about-the-covid-19-aftermath/.

182   Peter S. Goodman, "Sweden has become the world's coronavirus cautionary tale," New York Times, July 7, 2020, https://www.seattletimes.com/business/sweden-has-become-the-worlds-cautionary-tale/.

As incredible as it may seem, "U.S. manufacturers shipped millions of dollars' worth of face masks and other protective medical equipment to China in January and February *with the encouragement of the federal government.*"[183] Furthermore, "the move underscores the Trump administration's failure to recognize and prepare for the growing pandemic threat. In those two months, the value of protective masks and related items exported from the United States to China grew more than 1,000 percent compared with the same time last year—from $1.4 to about $17.6 million." On Feb 26—when total deaths had reached 2,770, nearly all in China—the Commerce Department published a flyer titled "CS China COVID procurement Service." It guided American firms on how to sell "critical medical products" to China and Hong Kong through Beijing's fast-tracked sales process."

On April 2, the U.S. government reversed course completely when President Donald Trump said his administration would invoke the Defense Production Act in a way that could have prevented 3M from selling masks to foreign customers, requiring the company to provide them to U.S. customers first. "We hit 3M hard today after seeing what they were doing with their masks," self-appointed "wartime president" Trump tweeted.

In the middle of July, President Trump first agreed that masks should be worn and did indeed wear one.

At this juncture it is too early to speculate about what will happen, and when, in the short term. History about the 1918 pandemic period, its aftermath, and the parallels between world wars and recessions is a major factor by which to gauge the mid- and long-term economic and social impact from the 2020 novel COVID-19 virus pandemic. It is not a stretch to say that the rapid recovery from the 1918 pandemic coupled with a short span of rejuvenation from WWI was a major factor contributing to the "Roaring 20s," during which the U.S. economy boomed. The economy grew for approximately seven years. But then came the dramatic crash of the stock market,

---

183 Desmond Butler, Juliet Eilperin, Tom Hamburger, and Jeff Stein, "U.S. sent millions of face masks to China early this year, ignoring pandemic warning signs," *Washington Post*, April 19, 2020, https://www.washingtonpost.com/health/us-sent-millions-of-face-masks-to-china-early-this-year-ignoring-pandemic-warning-signs/2020/04/18/aaccf54a-7ff5-11ea-8013-1b6da0e4a2b7_story.html.

which led to the Great Depression that lasted from August 1929 to June 1938. Notably for the 2020 situation, unemployment remained above 10 percent until 1941, when the U.S. entered World War II. The United States then suffered a number of social and economic ups and downs, culminating with the Great Recession in December 2007. That lasted 18 months, until June 2009.

The eleventh anniversary of the longest running stock bull market in history began on March 9, 2009, and coincidentally ended on March 9, 2020, approximately two months into the pandemic. The principal reasons for its end were fears of a global economic slowdown resulting from a U.S.–China trade war and rising U.S. interest rates. Both can be blamed on the current U.S. administration. All the above history offers clear indications that the next several years are going to be difficult, perilous, and virtually impossible to foretell.

The IMF forecast in January 2020 was that 160 nations would enjoy positive income growth. At the end of June, the IMF expected the global economy to shrink 4.9 percent this year. In brief, fear and trepidation about the mid-term and long-term future of the U.S economy are justified. So, can the economy and many people's lives be any worse? Sadly, yes. Many experts maintain that the latter part of this decade marks the beginning of massive joblessness. In addition to the jobs lost from the current world financial woes and the pandemic, artificial intelligence developers will soldier on in efforts to reduce business costs—the harbinger of still further job losses.

JOBS SLOW TO RE-EMERGE

A 2013 Gallup Poll conducted near the official end of the Great Recession revealed that although 30 percent of workers in the United States "were engaged, or involved in, enthusiastic about and committed to their workplace," 50 percent were not engaged.[184] The remaining 20 percent were "actively disengaged" and hated 'going to work, thus undermining their companies with their attitude. In brief, the poll's analysts

---

184   Richard Lopez, "Most hate their jobs, check out mentally at work," *Seattle Times*, June 18, 2013, http://seattletimes.com/html/nationworld/2021217458_gallupworkxml.html.

concluded that 70 percent of workers at that time hated their jobs or were mentally checked out. The result was an incredible economic impact on the nation, estimated at $550 billion in lost economic activity annually.

In 2017, eight years after the official end of the Great Recession, 51 percent of employees said they were very or somewhat satisfied with their jobs. They had good reason to be. There was full employment, and layoffs were at near-record lows. Yet there are gloomy sides. Only 52 percent of workers said they felt safe from a layoff, and "minimal raises and lean staffing has led them to lower their expectations on decent-paying jobs." In 1993, 73 percent of workers were confident they would be able to afford a comfortable retirement. In 2017, only 60 percent felt that way. "The traditional bond between employer and employee—in which companies provided job and retirement security in exchange for hard work and loyalty—has eroded so much that young workers today 'don't even know what they're missing,'" explained Rick Wartzman.[185] The end of the bull market, combined with the onslaught of the pandemic, has led to an image concordant with the 2013 Gallup Polls findings.

These projections indicate that the vast majority of ordinary American people will become even worse off. Potential jobs coming out of the pandemic will be slow to emerge, and a substantial portion will be gig ones, rather than stable, full-time ones. The overall scenario is that, on top of jobs lost during the COVID-19 epidemic and the impending recession, the most devastation will be caused by AI creations.

The saddest part and essence of this message is that "joblessness is a personal crisis because work is a spiritual event," according to Peggy Noonan. "Work gives us *purpose*, stability, integration, shared mission. And so to be unable to work—is a kind of catastrophe for a human being." [186] (Emphasis mine). The heart of the matter is that the threat of all these conditions poses a supreme challenge for lawmakers

---

185 Lauren Weber, "As Workers Expect Less, Job Satisfaction Rises," *Wall Street Journal*, September 1, 2017, https://www.wsj.com/articles/americans-are-happier-at-work-but-expect-a-lot-less-1504258201.

186 Peggy Noonan, "Work and the American Character," *Wall Street Journal*, August 30, 2013.

that are supposed to be defenders of our republican democratic system. So, specifically, from an American viewpoint, will our leaders in federal government take happiness and jobs as their measurement of social progress during these coming tough times for the masses? Do you achieve happiness through artificial intelligence? Does your life have purpose? Can you recalibrate as necessary?

11

THE ROLE OF HUMAN RIGHTS IN PUBLIC POLICY

She wondered with despair how she could have ever hated that life.

Because Zora once told her, you want something better, something more than mere happiness.

And where had that gotten her? What kind of arrogant bitch could claim that there was anything more important than happiness? What kind of a fool could believe such a twisted philosophy?

This fool, this one here, sitting huddled on a stranger's terrace, unable to speak a word to her hostess. This was how you ended up alone.

—Olen Steinhauer, *The Cairo Affair*

Happiness matters to humans because it is a principal contributor to our quality of life and integral to who we, as individuals and a society, are, and to what we would like our future to reflect. Happiness is part of the systems we refer to on a daily basis, such as when we inquire "How's it going?" Is it possible that people are not cognizant of being happy? One day I was checking some books out of the library. The nice, polite young lady asked, in a friendly fashion, "How is your day going?" I answered, "Super, operating at 110 percent, and quite happy. How about you—happy?" She paused, looked up, and contemplated long about it, as if it was something she had not thought about much. She answered, "Well, I'm not sad, so I guess I'm probably happy." So, how about you? Happy? Things going your way?

FREEDOM FROM VERSUS FREEDOM TO CARRY OUT RESEARCH

There is one more element that is germane to conditions surrounding the precautionary principle: human rights' role in scientific research and humankind's happiness. As might be expected, the scientific research issue is complicated, as Silja Voeneky points out in her chapter "Human Rights and the Legitimate Governance of Existential and Global Catastrophic Risks" in the 2018 book *Human Rights, Democracy, and Legitimacy in a World of Disorder.*[187] Apart from her discussion of the existential and global catastrophic risks, Voeneky argues that the existing human rights framework has so far been left aside as a (potential) important basis and starting point for a legitimate governance regime.

HUMAN RIGHTS' ROLE IN SCIENTIFIC RESEARCH

Now it's time for some critical thinking. It is well known that scientists value *freedom to* carry out research that can lead to "existential and global catastrophic risks." But, can they be controlled? Silja Voeneky affirms that freedom of research is not only a justified (i.e. moral or ethical) value—it is also a legally binding human right.[188] There is a shared view that the freedom of research is entailed in the right of freedom of thought and freedom of expression in international human rights treaties. But don't some parties also have *freedom from* those scientists' research?

The answer is yes, as Voeneky goes on to write: "To protect freedom of research as a human right does not mean that this freedom is absolute. According to legal international human rights, the protection of the life and health of human beings are—inter alia [among other things]—legitimate aims that can justify proportional limitations[189] of the right of freedom of science."

---

187   Silja Voeneky and Gerald L. Neuman, eds., *Human Rights, Democracy, and Legitimacy in a World of Disorder* (Cambridge: Cambridge University Press, 2018).

188   Silja Voeneky and Gerald L. Neuman, eds., *Human Rights, Democracy, and Legitimacy in a World of Disorder* (Cambridge: Cambridge University Press, 2018).

189   What proportionality means is linked to the risks and benefits one can reasonably anticipate.

Voeneky explains that international treaties "obligate states not only to respect, but also to protect the fundamental rights of individuals.... They state that States parties are obligated by international human rights treaties to take appropriate [legal] measures to protect the life of individuals."[190] In her view, "this duty includes a duty to protect the life of individuals against risk in low or unknown probability scenarios—which means that *no actual or direct threat* for a protected right exists—as long as there are risks of an existential or globally catastrophic nature."[191]

The UN Human Rights Committee's 2017 draft states that "the duty to protect the right to life by law also includes an obligation for States parties to take appropriate legal measures in order to protect life from all *foreseeable threats*, including threats emanating from private persons and entities."[192] The 2017 draft comment spells out similarly, later on, that "States parties are thus under a due diligence obligation to undertake reasonable positive measures, which do not impose on them impossible or disproportionate burdens, in *response to foreseeable threats to life* originating from private persons and entities, whose conduct is not attributable to the State."[193]

Silja Voeneky sheds light on limitations of scientific researchers, the wealthy, and government's apparent power to have control over development of superintelligence and radicalized humanoids, writing: "It does not seem to be a disproportionate limitation of science or technological progress to lay down a rule that there is a burden of proof for those who fund science (or for scientists) to show that there are more benefits than risks if there is plausibility for an existential or global catastrophic risk."[194]

She declares that because of the global dimension of the risks, it seems more plausible to argue that we need a global consensus to solve the problem of *freedom to*, as well as to determine whether to allow or prohibit certain experiments or

---

190 Ibid., 155.

191 Ibid., 155-156, emphasis in original.

192 As cited in Voeneky and Neuman, 156, emphasis in original.

193 As cited in Voeneky and Neuman, 157, emphasis in original.

194 Voeneky and Neuman, 161.

techniques. In simple terms, all of this means that regulations, controls, and bans can legally be put in place by relevant governmental authorities and others in the same field of scientific endeavor.

Human Rights' Role in Happiness

The UN Human Rights Committee's 2017 draft stresses that the right to life "concerns the entitlement of individuals to be free from acts and omissions intended or expected to cause their unnatural or premature death...." This draft comment holds that the duty to protect the right of society as a whole "implies that States parties must establish a legal framework to ensure the full enjoyment of the right to life by all individuals."[195] To me, the term enjoyment of the right to life in this 2017 draft translates into a social responsibility to protect happiness and quality of life.

We do live in a stressful, rapidly changing new reality. So, what is required to achieve happiness? Evaluation of what makes us happy and how to measure it is of ultimate importance because our future can be a source of happiness or of great sadness, depending on choices made about calamitous coming events. A rhetorical question: would Americans, and humans the world over, be happier, or really care if superintelligence and radically enhanced humanoids directly or indirectly impinged on our lives?

The interest and concern about individual and societal happiness is much greater than might be surmised, partly because happiness is a complex value. Thus, it is fascinating to many researchers since it is based on so many different things. For example, happiness includes allied concepts such as well-being and quality of life. As such, there are many ways happiness can be measured and, of course, there are cultural differences between peoples.[196] Most researchers focus on measuring degrees of happiness through large database surveys that measure life as a whole rather than current feelings. That

---

195 Ibid., 156.

196 The term "peoples" is often used at a traditional level because people of similar ethnic backgrounds often live across contemporary national borders.

is because those survey results can yield summations on which life conditions and circumstances are important for subjective well-being and happiness.

The *World Happiness Report*: A Vital Critical Thinking Tool

The first World Happiness Report (WHR), [197] a must reference for anyone concerned about our earth and who we can be, was commissioned for the April 2, 2012 United Nations Conference on Happiness (mandated by the UN General Assembly) and released on that date.[198] It is a vital critical thinking tool, for it reveals the importance and significance of the need to contemplate both America's future and our worldwide collective destiny.

Key conclusions from the first *World Happiness Report* are:
- Mental health is the biggest factor affecting happiness in any country. Yet only a quarter of mentally ill people get treatment for their condition in advanced countries and fewer in the developing nations.
- Stable family life and enduring marriages are important for the happiness of parents and children.
- Behaving well makes people happier.
- Women in advanced countries are happier than men, while the position in poorer countries is mixed.
- Happiness is lowest in middle age. ...
- Unemployment causes as much unhappiness as bereavement or separation. At work, security and good relationships do more for job satisfaction than high pay and convenient hours.
- A society's happiness only increases with income up to a point, as there is diminishing marginal utility of income.
- Social factors like absence of corruption, the degree of personal freedom, and social support are more important than income.

---

[197] John Helliwell, Richard Layard, and Jeffery Sachs, *World Happiness Report*, The Earth Institute, Columbia University, 2012. http://www.earth.columbia.edu/sitefiles/file/SachsWriting/2012/WorldHappinessReport.pdf.

[198] The Earth Institute, Columbia University, "First World Happiness Report Launched at the United Nations," April 2, 2012, http://www.earth.columbia.edu/articles/view/2960.

- On average, the world has become a happier place in the last 30 years [but not in the United States].

The *World Happiness Report 2020*[199] uses data drawn from the Gallup World Poll—a set of nationally representative surveys undertaken in more than 160 countries in over 140 languages. The ranking of happiness 2017–2019 report revealed that four closely ranking top countries, Finland, Denmark, Switzerland, and Iceland, rank high on all the main factors found to support happiness: caring, freedom, generosity, honesty, health, income, and good governance. All of the other countries in the top ten—Norway, the Netherlands, Sweden, New Zealand, Austria, and Luxembourg—also have high values in all six of the key variables used to explain happiness differences among countries and throughout time. These are income, healthy life expectancy, having someone to count on in times of trouble, generosity, freedom, and trust, with the top ten countries measured by the absence of corruption in business and government.

The United States is a testimony to reduced happiness in factors that are both social and personal. In 2020, America came in eighteenth in the world, for two main reasons: declining social support and increased corruption. Most important for this book is that the role of values and madcap development and adoption of technology are not directly addressed in the WHR. However, even casual perusal of the document leads to the inescapable conclusion that happiness surveys are centerpieces for evaluation about what is really happening to our society. Furthermore, they are essential components of improving our *life satisfaction* (used as a synonym for the term *happiness*).

Nicholas Kristof highlighted, in 2015, the increasing importance of quality of life replacing economic and plutocracy-driven indicators as key for policymakers to discern how and toward what to guide America. The reason? Ultimately, social progress is a critical measure of how a country is serving its people. Kristof declared, "As an American, what saddens me is also that our political system seems unable to rise to the

---

199   John Helliwell, Richard Layard, Jeffrey Sachs, Jan-Emmanuel De Neve, Haifang Huang, and Shun Wang, *World Happiness Report 2017* (New York: Sustainable Development Solutions Network, 2017).

challenges."[200] Some major reasons for that include the reality that the values of trust and civility in America are in decline. In addition, when policies and vision about our destiny are factored in, there is reason for fear, given that confidence in government is at an all-time low.

One vital key to understanding life satisfaction is that humans are social animals. As a consequence, meeting social norms and having a sense of belonging to the community are very important for human happiness. This factor is important considering the growing disconnect individuals in America are experiencing in face-to-face interaction as a result of rapid adoption of a wide variety of gadgets as part of the race to technologize our lives. However, strangely, the stay-at-home regulations that arrived with the 2020 pandemic have brought about closeness within families; they have also given rise to novel ways of conducting and even improving the education system. Ironically, and in an indirect way, the pandemic has brought about what hopefully will be social progress through the fight to eradicate racism.

HAPPINESS AS A PUBLIC POLICY OBJECTIVE

A substantial portion of the first *World Happiness Report* is devoted to overturning the traditional belief that happiness is only in the eyes of the beholder, something to be pursued individually rather than a matter of public policy. It can be added that such a belief is particularly strong in the moral ideologies of conservatives and libertarians, just as the belief in rugged individualism is largely a sacred value to them. Thus, the prevailing view by conservatives is that happiness seems far too subjective and too vague to serve as a touchstone for national goals, much less policy content.

Contrary to this conservative view, the authors of the first WHR argued that happiness can be objectively measured. They arrived at that finding through evidence gleaned from the emerging scientific study of happiness by psychologists,

---

[200] Nicholas Kristof, "Enjoying the low life?" *New York Times*, April 9, 2015, http://www.nytimes.com/2015/04/09/opinion/nicholas-kristof-enjoy ing-the-low-life.html?_r=0.

economists, pollsters, sociologists, and others who treat happiness as though it were a measurable, objective study. They also concluded that happiness can be assessed when correlated with brain functions and related to characteristics of individuals and societies. Additionally, they surmised that if it is agreed upon that societies should foster the (ultimate) happiness of citizens, it is logical that visions of a good society should be based on subjective life satisfaction information, not just the goal of increasing GDP.[201] The conclusion of the WHR is that "happiness depends on a huge range of influences, many of which can be influenced by government policy. ... The calculations confirm the powerful effect of many variables other than income."[202]

Determination of One's Own Happiness

There is one more aspect of happiness: quality of life. The term is not specifically mentioned in surveys but is embodied in many of the questions about such things as life satisfaction. Actually, quality of life is similar to happiness in the sense that we (mostly) know it when we have it or do not have it. The University of Toronto's Quality of Life Research Unit states that from their perspective, "the ultimate goal of quality of life study and its subsequent applications is to enable people to live quality lives—lives that are both meaningful and enjoyed."[203] There is considerable use of the term "quality of life" in politics and policy goals. However, it actually is a personal issue, because it means so many different things to different people. For some, it is an economic measure. For others, it is about achieving aspirations and missions in life.

Dr. Abraham Maslow developed a scientifically articulated, viable theory about the *purpose* of each individual's life.[204]

---

201   Two excellent books that that follow that same theme: Derek Bok, *The Politics of Happiness: What Government Can Learn from the New Research on Well-Being* (Princeton, NJ: Princeton University Press, 2010), and Carol Graham, *Happiness Around the World: The Paradox of Happy Peasants and Miserable Millionaires* (New York: Oxford University Press, 2010).

202   *World Happiness Report*, 2020.

203   Quality of Life Research Unit, "Welcome," n.d., http://sites.utoronto.ca/qol/.

204   Abraham H. Maslow, *The Further Reaches of Human Nature* (New York: Viking Press, 1971).

The founder of humanistic psychology determined there are five distinct, hierarchical levels of human needs. Generally speaking, the higher needs emerge as the needs in the lower tiers are satisfied. Survival and meeting the "basic (physiological) needs"—for air, food, water, warmth, and rest—is the primary concern of all living organisms. The next higher level is meeting "security needs," including "protection from elements, security, order, law, stability, freedom from fear."[205] Once a being is secure, relationships of friendship and love are sought, as is affiliation with a group. This constitutes the third tier of Maslow's human needs: for "love and belonging." Once those needs are met, we turn our attention inward, developing esteem for ourselves, and outward, to attain recognition from others, termed "ego (esteem) needs, the fourth level."

One of Maslow's most important discoveries is that, once we are free from all stress and can actualize our existence and resonate in harmony with our own environment, we may realize our sense of peace, in which personal potential, self-fulfillment, and seeking personal growth leads to "peak experiences." This fifth level is a desire "to become everything one is capable of becoming (Maslow, 1987, p. 64)."[206]

Maslow pointed out that most people never have a peak experience in their lives, as they are too engaged in meeting the lower four categories of needs—especially economic needs. Another conclusion in our brave new era is that today, nearly every aspect of our lives has become "gadgetized." A distressing example of our tech-oriented lives is that when couples or friends get together, a significant proportion of them focus on a gadget when they are together rather than conversing with each other.

Similar to Maslow's determinations, I argue that quality of life and happiness, whether recognized or not, can be attained only when one has the illusion of being in control of one's life. The feeling that one is in control is not like controlling other people. It means being free of troublesome things that make one unhappy, such as frustration about being harried and unable to meet daily needs and challenges. It means not feeling

---

205   Saul McLeod, "Maslow's Hierarchy of Needs," March 20, 2020, https://www.simplypsychology.org/maslow.html.

206   Ibid.

overwhelmed. It means being in control of events that impinge on one's time and tranquility. In brief, the feeling of being in control evokes the same feelings of serenity, acceptance, and joy that the heightened sense of awareness of mindfulness can. The notion of accepting one's feelings, thoughts, and bodily sensations can also be used as a therapeutic technique to deepen one's sense of *purpose*.

Lyle Simpson (no relation to the author), in his short publication *Why Was I Born?*, critically evaluated the *purpose* of life in light of Maslow's hierarchy of needs.[207] Simpson argues that life is meaningful to the extent that we share in happiness, and that the world becomes a better place because we have lived. What he means is that just doing nice things for others as a mission in life is a way to improve one's own little world. That notion holds for collective happiness as well. In brief, if you want greater happiness, it is up to you to make the effort to get it. You can't just sit around waiting for it to be handed to you.

Choice for a Posthuman Condition

The question of gaining happiness from any kind of artificial intelligence is an intriguing one.

Let's take a rather extreme example about *purpose* in individual lives and enhancements. Imagine a person whose basic and security needs are met but who lacks friendship and love, the third level in Maslow's theory about happiness. That person seems to have no reason for living each day in their current unenhanced conditions and finds no *purpose* to their current life. Then assume that individual becomes acquainted with transhumanist philosophies and begins to wonder, "Who am I anyway? What do I want out of life? What's wrong with my life now? I don't even seem to have any *purpose* in life. So, why not explore life through radical enhancements, the ones that some transhumanists aspire to? It sounds like nirvana. That might be just the ticket."

That individual finally decides to do it, and seeks out and acquires enhancements. Perhaps that person is truly in

---

[207] Lyle L. Simpson, *Why Was I Born? What Is My Purpose of Being Here? A Humanistic View of Life* (Washington, DC: The Humanist Press, Third Edition, 2013). Emphasis mine.

a Shangri-La, free from all stress, harmoniously living in that environment, and winds up bouncing from one peak experience to the next. But maybe that endowed person's transformation turns into a perpetual nightmare. What then? Will there be a reversal process for enhancements that would allow a person to regain their original brain and personality? Suppose you decide to take the leap. You wind up in a posthuman condition. Your new life is not what you had been promised. Would you wonder with despair how you could have ever hated your pre-enhanced life? Would you feel like a fool, wondering how you ended up alone?

Do you believe that we have the right to control our own destinies? How about the populace being the choice makers of our futures? Can citizens be trusted to make rational choices? Why should action be taken? Why not just let events unfold?

12

THE WAY FORWARD

"The only thing necessary for the triumph of evil is for good men to do nothing."

This quote has been attributed to Edmund Burke and was included by John F. Kennedy in a speech in 1961. Its earliest form was by John Stuart Mill, who said in 1867: "Bad men need nothing more to compass their ends, than that good men should look on and do nothing."

Critical questions are in order. Are those around the world who create and develop the technologies that will dramatically affect our destinies truly be concerned about life satisfaction, quality of life, or whatever else happiness is termed? Will we control intelligent machines, or will they control us? What would it mean to be human in an age of AI when intelligent machines coexist with or replace us?

THE RIGHT TO CONTROL OUR DESTINY

A day of reckoning will come faster than we can imagine. As an inducement to mull over that fact, I repeat Ray Kurzweil's thesis that singularity should be advanced as soon as possible because "waking up the universe, and then intelligently deciding its fate by infusing it with our human intelligence in its nonbiological form, is our destiny."[208] Is this a scenario you find attractive? The reality is that if nothing changes regarding regulations, controls, or bans over the next three decades, you and humanity will not be safe. This is not science fiction. A number of well-

---
208   Ray Kurzweil, *How to Create a Mind: The Secret of Human Thought Revealed* (New York: Viking Penguin, 2012), 282.

known futurist authors on technology reckon that this is the impact AI could have on humankind, as we know our species.

This discussion about humankind's destiny has a counterpart dilemma. Scientists warn of mass extinctions in the next two decades as humans continue to encroach on nature. Gerardo Ceballos, a well-known ecologist, and his colleagues warn of a cascading series of impacts—including more frequent occurrences of new diseases and pandemics. Ceballos states that there is no way this can be continued without putting the whole of humanity in danger. His view is that this immediate problem has to be solved in the not-too-distant future, and in an interview with the *New York Times*, he notes: "Making these changes will require electing leaders who prioritize the environment, redistributing resources and slowing human population growth."[209] Unfortunately, many who are aware of the situation may simply feel the loss is not consequential. Ceballos adds, "People say, what the hell of a difference does it make to me?"

CHOICE MAKERS OF OUR FUTURE

Should the populace be content to relegate our future and happiness to the whims of those in positions of power regarding regulations on superintelligence and radically enhanced humanoids?

Unfortunately, at this point, trust in government to do the right thing and stand up to the elites is so low that the triumvirate has a clear advantage. This is because they have time on their side to advance technologies if no strong opposition emerges. So, in reality, what can be done by the citizens to ensure appropriate measures are taken at prudent, well-judged times to avoid catastrophes from superintelligence and radical humanoid development?

Tragically, a problem exists that begs for critical thinking. What shall we do if the big three take a hardened stance against regulations, controls, or bans on AI technologies that have existential risk? The quick answer is that our government would

---

209 Rachel Nuwer, *New York Times*, June 1, 2020, https://www.nytimes.com/2020/06/01/science/mass-extinctions-are-accelerating-scientists-report.html.

have to break loose, stand up, and act against what is morally wrong. An equally short answer is use of the *precautionary principle* as the logical legal course of action. Why? Because without it, the loss of humankind, as we know it, is the price we'd have to pay. It will be a struggle for the voters to effectively rise up against powerful forces they know little about, because the three entities don't want to be disturbed or give up power. A serious complication is that neither the public nor our American government thinks beyond the immediacy of the near future.

Gerald Baker has pinpointed this stumbling block about America: "For all the name-calling and blame-apportioning we are now doing, the evidence of much of history is that the failure to respond adequately to a looming threat is embedded in our nature. Expecting us to respond to an impending challenge by taking the most extreme measures to prevent it is expecting something we have rarely done—something ahistoric and virtually unhuman. …We never act until it's too late, mainly because most of us don't believe the threat is real until it is."[210] Mull over the 2020 pandemic. Must it take a life-threatening crisis like this to provoke/elicit/trigger a call to action?

A Call to Action

Each new generation wants to explore and feel adventurous. That is natural. The difference from earlier generations shared by many of the two younger sets, Generation Z, born 1997–2012, and the Post-Zers, is that rapid change is a desirable and constant element in their lives. It is not an exaggeration to argue that the younger generation worships change and views it as a definition of progress. Thus, continual release of new technological creations is high on their list of priorities. What is also clearly discernible is the acceptance of the notion that technologies also control much of their lives. Another grave downside is that many don't realize that progress is the process of improving or getting nearer to achieving or accomplishing

---

210   Gerald Baker, "We Only Fight Threats When They're Upon Us," *Wall Street Journal*, April 4-5, 2020, https://www.wsj.com/articles/we-only-ever-fight-threats-when-theyre-upon-us-11585932107.

something beneficial for the future of society. But cheer up! Many do find *purpose* in their lives and will fight for the good of humankind. So where does that leave us? Can citizens indeed be trusted to make rational choices?

Edith Hall writes that Aristotle goes so far as to say, "At the beginning of *Eudemian Ethics,* Aristotle quotes a line of wisdom literature inscribed on an ancient stone on the sacred island of Delos. It is proclaimed that the three best things in life are, 'Justice, Health and Achieving One's Desires.' Aristotle trenchantly disagrees. According to him, the ultimate goal in life is simply, happiness, which means finding purpose in order to realize your potential and working on your behavior to become the best version of yourself. You are your own moral agent, but act in an interconnected world where partnerships with other people are of great significance."[211]

Personally, I do hold a positive view and have reason to believe an overwhelming percentage of the population at all levels can be trusted to make decisions on a destiny focused on the wellbeing of humankind as we know it. However, in our case, they would have to be properly informed about the advantageous and detrimental aspects of artificial intelligence. That means beginning soon. It means teachers at all levels and our government have a profound duty to continually explain the pros and cons of artificial intelligence—as this book suggests, particularly regarding superintelligence and radically enhanced humanoids.

On a positive note, David Brooks wrote in his July 4, 2019 column "Will Gen-Z Save the World?" that "it's often uncomfortable and over the top. But we're lucky to have a rebellion against boomer quietism and moral miniaturization. The young zealots may burn us all in the flames of their auto-da-fe [public penance], but it's better than living in the Boomer society marked by loneliness and quiet despair."[212]

One day I happened to be chatting with a neighbor who was taking care of his grandchild, a lively four-year-old girl with trusting brown eyes who obviously adored her grandfather.

---

[211] Edith Hall, *Aristotle's Way: How Ancient Wisdom Can Change Your Life.* (New York: Penguin Press, 2019). [quote on page 26].

[212] David Brooks, "Will Gen-Z Save the World?" *New York Times*, July 4, 2019, https://www.nytimes.com/2019/07/04/opinion/gen-z-boomers.html.

My friend asked me, "Jim, do you have grandchildren?" I replied, "No." Out of natural curiosity, he asked me why, as he saw how much I was enjoying his little one. "I imagine you miss having the little tykes and watching them grow up. This grandchild forms the greatest happiness of my life. I know you have three grown children, two sons and a daughter." "Well," I answered, "like your children, they are doing well in their lives. We have just let them be independent thinkers and make their own decisions about the children aspect. They have well-thought-out reasons for not having children, like wanting to devote their lives to jobs and spouses. Regrettably, there is also fear of the future, and what it bodes from a number of angles." My quick calculation was that this energetic four-year-old girl born in 2015 will be fifteen in 2030. That is the time at which Verner Vinge believed singularity could take place. She will be thirty in 2045, when Ray Kurzweil projected there will be machines whose intelligence is greater than that of humans, and the emergence of singularity.

What I have provided is an early warning, an explanation, and evidence about the technologies that will affect your life, your children's lives, and the lives of others on this planet. Now, dear reader, it is up to you to keep this conversation going. The bottom line on decisions about humanity, as we know it, comes down to three broad choices: (1) just let events unfold, (2) submit to the adventure of unfettered artificial intelligence development, or (3) take action to judiciously place controls on the existential risks detailed in this book. I have my lovely wife, three remarkable children, and friends all over the world with children. I know my choice. I emphatically vote to take action.

The cause of freedom is not the cause of a race or a sect, a party or a class—it is the cause of humankind, the very birthright of humanity.

This quote is attributed to Anna Julia Cooper, who was born into slavery in 1858 and died in 1964 at the age of 105. It first appeared on United States passports in 2012 and has remained there ever since.

Building Blocks for Humanoids

My objective in developing these building blocks is to provide something for everyone thirsting for more knowledge about humanoid creation and development, to make the content informative for people at all levels of scientific knowledge on the topic and simultaneously a pleasure to read. I know readers of this book vary from those with minimal interest in how humanoids will be developed, to those in the multitude of scientific fields who have a deep interest in certain aspects of the innovations and creations. To that end, I encourage both types of readers to enjoy critical thinking about three key topics: how the humanoid timetable might change, at what point in it humanoids could become an existential risk, and whether radically enhanced humanoids should be controlled or banned—and if so, when and how.

Part 1. The Path to Humanoid Robot and Humanoid Development
- Close Coupling of Humans and Machines
- Mobility, Vision, and Appearance of Robots
- Body Parts Regeneration by 3-D Printing
- Power to Operate and Speech
- Transplants of Body Parts
- Reproduction: Womb Transfer and Surrogate Babies
- Embryo Transfer

Part 2. Creating Humanoids as Living Organisms

- Some Genomics Are Truly Unnerving
- Cloning
- Cloning Humans
- Thinking, Emotions and Brains for Being a Living Organism
- Should Humanoids Be Given Free Will?
- Humanoids as a New Species

Part 1

THE PATH TO HUMANOID ROBOT AND HUMANOID DEVELOPMENT

Man–computer symbiosis is an expected development in cooperative interaction between men and electronic computers. It will involve very close coupling between the human and the electronic members of the partnership.

—J. C. R. Licklider,
Ire Transactions on Human Factors in Electronics (March 1960)

Development of functional humanoids as near equals to humans is rapidly advancing. Example: evidence suggests that robotized humanoids will have sufficient mobility to interact with humans, if the creators want them to, in the not-too-distant future. A team of Harvard University researchers announced the revolutionary first autonomous, soft, untethered, and 3-D printed hybrid between an octopus-like creature and a robot in mid-2016. Each of the functional components required—fuel storage, power, and actuation—is included in "these curious creatures, [which] can perform incredible feats of strength and dexterity with no internal skeleton." This squishy creature is cheap to print and could pave the way for a new generation of such machines.[213] It goes without saying that this innovation can speed up the race to develop advanced humanoids. Now onward to my favorite topic: symbiosis, perhaps because it is an opportunity to tell the tale of my dog Sam.

---

213   Leah Burrows, "The first autonomous, entirely soft robot," *Harvard Gazette*, http://news.harvard.edu/gazette/story/2016/08/the-first-autonomous-entirely-soft-robot/

Close Coupling of Humans and Machines

The field of biology is where we obtain background to evaluate pathways into and through symbiosis to creations beyond single-celled, or unicellular, organisms, the smallest contiguous unit of life.[214] Organisms can grow, respond to stimuli, reproduce, and, through evolution, adapt to their environment in successive generations. The 1879 definition of symbiosis by Heinrich Anton de Bary as the living together of unlike organisms now includes all species. Another commonly used definition of symbiosis is simply the close and often long-term interaction between two or more biological species.

There are multiple ways to categorize symbiosis. For the sake of brevity, I will address only one of those systems. Symbiosis can be viewed as being on a continuum between antagonistic and cooperative symbiotic relationships. *Antagonistic* relationships occur between hosts and parasites or pathogens. In the antagonistic relationship of parasitism, the parasite generally gains, while the host is harmed. *Cooperative* relationships are on the opposite end of the spectrum. One such cooperative relationship is known as *mutualism*. Mutualism is where the two parties both benefit. *Commensalism* still lies on the cooperative side of the symbiosis spectrum, but a bit farther away from the end than mutualism. Commensalism, a biological term that describes the relationship between two living organisms where one benefits and the other is not significantly harmed or helped, is an important concept in the future evolution of mutualism between humans and machines.

For example, in commensalism transportation by one organism of another or living together in housing is a type of symbiosis. Another is known as *phoresy*, "a type of biological hitch-hiking" whereby transportation is provided by one organism to a second organism.[215] A third example is two organisms living together in shared housing, or one organism taking up residence in the abandoned dwelling of another. It is

---

214 I do hope that biologists do not take umbrage for my impinging on their purview in symbiosis. I am well aware of journals such as *Symbiosis* published by Springer and other sources. The objective in this chapter is to at least add to work, ideas, and interests of others on this topic, one that I take very seriously.

215 Encyclopedia.com, s.v. "Commensalism," accessed September 17, 2020, https://www.encyclopedia.com/science-and-technology/biology-and-genetics.

basically a one-sided symbiosis. Commensalism is an important concept in the future evolution of mutualistic symbiosis between humans and machines.

Some organisms are *obligate*, meaning they depend on another for survival. Others are termed *facultative*, meaning they can live with another organism, but do not have to. Another vital distinction is the classification by physical attachment. Those that have bodily union are termed *conjunctive*, while those in which there is not a union are *disjunctive*.

A common contemporary dictionary definition of mutualism is the relationship between two different living creatures that live close together and depend on each other in particular ways, each getting benefits from the other. The relationships in mutualism may be either obligate for both species, obligate for one but facultative for the other, or facultative for both. Biologists generally have focused on organisms that are extremely or relatively small. However, the term also applies to mutualisms between creatures that are relatively large, such as the goby fish, and those that are smaller, such as the shrimp.. These two creatures sometimes live together; the blind shrimp digs the hole, and the fish alerts it when there is danger so it can re-enter the hole. Most land plants and land ecosystems rely on mutualisms between the plants.

From a strictly biological perspective of mutualism, the relationship between humans and machines at present is not a symbiotic one. However, the increasingly close working relationships between humans and robots is symbiotic within the popular definition of symbiosis as a relationship between people, companies, and so forth that is to the advantage of both. That is the meaning Licklider had in mind in his 1960 publication about man-computer symbiosis.[216]

The use of words and terms is continually changing. For our purposes, think about *cobots*, robots in a factory that work side by side with humans on specified tasks. That interaction is a symbiotic relationship between humans and machines. Thus, a newer use of the term mutualism is warranted in the symbiotic development of humanoids to mean the way two entities, one human and the other a machine, exist in a relationship in

---

216   J. C. R. Licklider, "Man-Computer Symbiosis," in *IRE Transactions on Human Factors in Electronics*, vol. HFE-1, no. 1 (March 1960): 4-11.

which individuals benefit from the interaction of humans and machines—and even some joining of the two.

The obsession with animal ownership in America provides an example of symbiotic mutualism. In this case, there are two distinct living species involved. Let me personalize a case that many readers can relate to. It exemplifies the condition in which many biologists restrict the definition of symbiosis to close mutual relationships and, in general, only to lifelong interactions involving close physical and biochemical contact.

I got Sam in Tucson as a puppy when I was discharged from the Navy in 1961. We were inseparable. He waited under my pickup while I was in college attending classes, went up with me on ranches where I worked as a cowboy, and was a great asset in dating. Initially, our relationship was facultative. Then, at some point, while my outward feelings remained facultative, Sam's relationship had changed to obligate, meaning he depended on me for survival.

At the end of my study for my master's degree, I began an international career and left Sam with my folks on their farm. One day, in faraway Paraguay, I received a letter saying that my parents' veterinarian had diagnosed Sam as dying of loneliness for me. Unknowingly, my facultative relationship had developed to the point where I felt obligated to send for him. I did so, and he recovered quickly. I did have to leave him with my folks in later years, but he remained in good condition and my visits were sufficient for him to be happy right to the end.

Mobility, Vision, and Appearance of Robots

Enhancing mobility in robots is a step in the symbiotic process of adding human attributes to robotic platforms. *Biorobotics*, in the AI field, takes inspiration from biological principles to design robots with mobility that approaches that of animals. The breathtaking cyborgization of humans (a *cyborg* is a complex organism that has advanced abilities resulting from augmentations and enhancements, particularly mechanical parts) is ample to demonstrate that developments in robot mobility have been dramatic. For example, although arms are relatively easy and inexpensive to create, legs are very expensive and complex because of mathematical issues related to adjusting a foot in a certain manner, or even leaning on a leg. However, robots are on the cusp of being able to climb stairs, and, given rapid advances in technology, it is inevitable that robots will have ability to adroitly get in and out of vehicles.

Now, about vision: the biggest challenge for robots is to make sense of all the data coming into their cameras, because they are clumsy. However, a technology to attach tiny lenses to a sheet of some flexible underlay, developed at the University of Wisconsin–Madison, alleviates that problem. These micro lenses, each about as big as the head of a pin, could, for example, cover a robot's head or body. Apart from mobility, robots that can see are a key to self-driving cars, projected to be in use in the early 2020s. They are likely to be used in a wide variety of industries; for example, food manufacturers combine AI software and advances in laser vision for tasks like slicing meats. The sensing and imaging market will grow dramatically to meet unfathomable demand as devices that mimic vision become even more varied and able to meet almost every need.

Computer makers have now used Intel's computer vision technology, dubbed RealSense, in devices such as laptops and tablets. Intel RealSense technology is a suite of "depth and tracking technologies designed to give machines and devices depth perception capabilities" that will enable them to "see" and understand the world. Intel RealSense technology is made of Vision Processors, Depth and Tracking Modules, and Depth Cameras. In practice, this technology allows users to scan objects and people in three dimensions and put an image in a computer game or to 3-D print a miniature model.[217]

These techniques also permit control of vending machines with a wave of a hand, and, with digital mirrors, allow people to try on clothes virtually. Lowly toothbrushes can now be delivered to hotel rooms by computer vision robots. Paris-based Blue Frog Robotics SAS has included computer vision so that "Buddy, a home robot with cartoonish eyes," can recognize family members in the kitchen.[218]

---

217   Wikipedia, s.v. "Intel RealSense," last updated August 20, 2020, https://en.wikipedia.org/wiki/Intel_RealSense.

218   Jack Nicas, "Why Your Gadgets Can Now 'See' in 3-D," *Wall Street Journal*, updated October 15, 2015, https://www.wsj.com/articles/more-devices-gain-3-d-vision-1444859629#:~:text=ParisbasedBlueFrogRobotics,aBlueFrogexecutiveestimated.

AI PREVAILS

Much more advanced, and one vital step in the shadowy race to develop advanced humanoids, Intel's RealSense computer vision technology frees up robots to "leave their stationary posts in factories and navigate the real world."[219] By 2016, a grocery billionaire had developed a symbiotic automation system that includes "autonomous robots that can travel untethered among storage racks in a distribution center."[220] In just two years, these robots have become the norm in Amazon and other companies' warehouses.[221] On the individual level, the RealSense technology can also guide customers to the plumbing section of a hardware store. Alas, there goes another job for humankind.

Deep learning technology, which is a statistical technique that enables computers to learn by processing huge amounts of data, has been one of the hottest investment targets for global tech giants for a half decade. One beneficial way to apply learning techniques for machines is by teaching them to recognize spoken words. Research is focused on a kind of self-study program for machines in which computers learn by themselves how to achieve a task, rather than be programmed solely with fixed rules. By 2015, the Japanese robot maker Fanuc Corp. had strengthened a tie-up with Preferred Networks, an artificial-intelligence venture, as part of its effort to develop industrial machines that can learn and repair themselves.

Another physical biotech example is creation of artificial human skin that the recipient will not reject, which has long been a goal in medically oriented research, especially for burn patients with chronic wounds. A positive side benefit is that this technology aids in taking the place of animals in testing products such as cosmetics. Ongoing research is directed toward development of waterproof and flexible qualities like a wide range of realistic skin tones and fully functional hair follicles and sweat glands akin to those of normal skin.

---

219   Ibid.

220   Robbie Whelan, "Fully Autonomous Robots: The Warehouse Workers of the Near Future," *Wall Street Journal*, September 20, 2016, http://www.wsj.com/articles/fully-autonomous-robots-the-warehouse-workers-of-the-near-future-1474383024.

221   Angel Gonzales, "Hands, Heads, Robots," *Seattle Times*, April 9, 2016, http://www.seattletimes.com/business/amazon/at-amazon-warehouses-humans-and-robots-are-in-sync/?utm_source=twitter&utm_medium=social&utm_campaign=article_left_1.1.

Artificial skin has been developed that "acts like it knows when it is being touched and sends out the news as if by telegraph."[222]

An immediate use of artificial human skin is in prostheses. A biotech goal is creation of a flexible electronic covering that, when connected to the body's nervous system, approximates the powers of real skin. It is obvious that these skin technologies will be of great benefit to humans. Examples abound of other innovations, such as "an invisible film," developed by scientists in 2016, "that can be painted on your skin and give it the elasticity of youth."[223] By 2017, developers reported they had "genetically modified stem cells to grow skin that they successfully grafted over nearly all of a child's body."[224] One implication is that apart from making wrinkles disappear on humans, we can speculate that advanced humanoid sex workers and their customers might benefit.

Body Parts Regeneration by 3-D Printing

Although perfection is in the future: "printing" using the building blocks of human biomaterial is well underway in multiple laboratories to create tissues of kidneys as well as of livers, with the latter being the most regenerative organ in the body. In 2018, Organovo, a company based in San Diego, developed a bioprinting process that takes cells from the patient or adult stem cells and turns them into printable bio-ink. From that, they "build up small sections of liver tissue" by layering biomaterial in "carefully calculated designs."[225]

Many experts do caution that "convincing the government

---

222   Daniel Akst, "Artificial Skin That Knows When It Gets Touched," *Wall Street Journal*, October 22, 2015, http://www.wsj.com/articles/artificial-skin-that-knows-when-it-gets-touched-1445532088.

223   Gina Kolata, "'Second Skin' May Reduce Wrinkles, Eyebags, Scientists Say," *New York Times*, May 9, 2016, https://www.nytimes.com/2016/05/10/health/second-skin-aging-wrinkles.html.

224   Ariana Eunjung Cha, "Genetically modified skin grown from stem cells saved a 7-year-old boy's life," *Denver Post*, November 8, 2017, http://www.denverpost.com/2017/11/08/gene-therapy-stem-cells-skin-saves-boy/.

225   Hasan Chowdhury, "Liver success holds promise of 3D organ printing," *Financial Times*, March 4, 2018, https://www.ft.com/content/67e-3ab88-f56f-11e7-a4c9-bbdefa4f210b.

of the safety and efficacy of implanting bioprinted tissues into people" is "one of the biggest challenges" for the industry. Nevertheless, as reporter Steve Johnson wrote, "Despite challenges, many people are encouraged by the impact 3-D printing is already having on health care, especially in robotic prosthetics. There are a lot of hurdles, but there is a lot that shows it is worth going forward."[226]

In an unusual twist on 3-D printing, although most efforts are concentrated on improving the quality of life of living people, scientists have developed a way to benefit the dead. The first procedure, announced in 2018, involved initially scanning the donor's face, in this case a person who was brain-dead.[227] The next part of the procedure was to remove the donor's face and attach it to the living person, a disfigured man. The last step was to attach the lifelike 3-D replica, taken from the scan, to the dead donor. Naturally, not everyone will volunteer to be a donor for such a procedure. For that reason, the medical community of surgeons and related scientists is continually excited about the next opportunity for unusual innovations—in this case, using a printed face. That conclusion reasonably leads to the question of how long it will be until human body parts are printed for humanoids. And speaking of that, how about using wiggly things to power humanoids?

Power to Operate and Speech

Humanoid robots and humanoids, like humans, must have power to operate. Several options can be thought up, such as electrical plug-ins for a battery designed for humanoids, much like for electric cars or tools, or filling up at a fuel station. However, sources like that can be bulky and inconvenient, and they don't fit the bill for a humanoid to be on par with humans. Food that is specifically developed for humanoid

---

[226] Steve Johnson, "Researchers aim to push 3D printing into living organs, tissue," *Seattle Times*, February 8, 2015, http://seattletimes.com/html/businesstechnology/2025649409_3dprintbodypartsxml.html.

[227] Andy Newman and Marc Santora, "For the Living, a Donated Face. For the Dead, a Lifelike Replacement," *New York Times*, January 5, 2018, https://www.nytimes.com/2018/01/05/nyregion/face-transplant-3-d-printed-mask-donor-nyu.html?register=google.

robots would be a desirable source, so 3-D printers have been developed by such companies as ChefJetPro and Foodini to print out "successive layers of edible material" intended to match ingredients in recipes.[228] That sounds fine. However, food for power brings a problem: waste elimination. This time, 3-D printers become an essential component in a technology to help solve both the power and excretory process issues.

The startling innovation of a 3-D printed hybrid between an octopus-like creature and a robot, with functional components including fuel storage, power, and actuation[229] was created by Harvard researchers in 2016. That feat was added to, in 2019, by researchers at Harvard and Cal Tech. The 3-D printed Rollbot, which can move and change shape in response to external stimuli such as temperature and light, is "paving the way for fully untethered, soft robots" that could lead the way to such robots' being used in medicine and industrial engineering.[230] It doesn't take much imagination to visualize eager developers creating humanoid robots with state-of-the-art mobility and powered by these creations. It goes without saying that this innovation can speed up the race to develop advanced humanoids.

Now on to speech. Our voices are one of our main communication tools, so natural talk is an essential attribute for humans to assist humans in discerning whether humanoids are on a par with them. In 2011, IBM's Watson competed on *Jeopardy!* against legendary champions Ken Jennings and Brad Rutter, and Watson won one million dollars! Apple introduced Siri, the first modern vertical assistant, a short time later. Amazon's Alexa followed in 2014. In 2016, researchers using the International Business Machines Corporation's Watson analytics system developed speech that seemed "very much

---

228   Michelle Lock, "3-D printing aims to rewrite the script on cooking and tech," Phys.org, February 11, 2015, http://phys.org/news/2015-02-d-aims-rewrite-script-cooking.html.

229   Leah Burrows, "The first autonomous, entirely soft robot," *Harvard Gazette*, August 24, 2016, http://news.harvard.edu/gazette/story/2016/08/the-first-autonomous-entirely-soft-robot/.

230   Leah Burrows, SEAS Communications, "Self-folding 'Rollbot' paves the way for fully untethered soft robots," August 21, 2019, https://wyss.harvard.edu/self-folding-rollbot-paves-the-way-for-fully-untethered-soft-robots.

like a normal conversation with a human being."[231]

Alexa first appeared in the UK and Germany, followed by India, in October of 2017. Teams of linguists, speech scientists, developers, and engineers worked together on Amazon's Alexa to use a blend of Hindi and English, which "she" "speaks with an unmistakably Indian accent."[232] Then, in 2018, Alexa, Siri, Cortina, Google Assistant, and other talking assistants, termed *chatbots*, began to support multiple apps, including messaging apps.[233] The crux of the matter is that we can expect normal conversation from humanoid robots by around 2025, when the first basic ones begin to mingle with humans.

Voice-controlled virtual assistants such as the Amazon Echo, and an increasing number of customer-care websites, are the latest signs that big technology companies believe our future involves talking to computers that can talk back. Paradoxically, because the voices sound almost human, a debate is on about whether, at a bare minimum, a bot should self-identify and answer truthfully when asked if it is a bot. However, this issue is insignificant to AI developers who consider speech to be one of the formidable challenges for the world knowledge problem. To them, endowing AI with mastery of aspects of conversation natural to humans like irony, ambiguity, sarcasm, laughter, and puns, is the foremost goal.

By 2018, some scientists were asking whether deep learning is really so deep after all, because there is no real intelligence there—just brute-force algorithms based on huge quantities of data. The problem is that although the software can instantly identify millions of words, heightening interest in using deep learning techniques for speech, there is another way. In 2018, the Allen Institute in Seattle reported it "would invest $125 million over three years in research to generate common-sense knowledge" in an initiative called Project Alexandra. Not to be

---

231   Melissa Korn, "Imagine Discovering that Your Teaching Assistant Really Is a Robot," *Wall Street Journal*, May 6, 2016, http://www.wsj.com/articles/if-your-teacher-sounds-like-a-robot-you-might-be-on-to-something-1462546621.

232   Saritha Rai, "Amazon Teaches Alexa to Speak Hinglish. Apple's Siri Is Next," *Bloomberg News*, October 30, 2017, https://www.bloomberg.com/news/articles/2017-10-30/amazon-teaches-alexa-to-speak-hinglish-apple-s-siri-is-next.

233   Alexandra Samuel, "What I Learned From Building My Own Chatbot," *Wall Street Journal*, April 29, 2018, https://www.wsj.com/articles/what-i-learned-from-building-my-own-chatbot-1525053780.

outdone, in the same year, the Pentagon proposed a five-year, high-risk project with total funding of $60 million on the same topic.[234]

The world's top research labs are rapidly improving computers' ability to understand and respond to natural language. Machines are getting better at analyzing documents, finding information, answering questions, and generating language of their own. The Allen Institute for Artificial Intelligence unveiled a system termed Aristo, for Aristotle, in September 2019. An eighth-grade science level test was passed comfortably, and Aristo answered 80 percent of the questions on the twelfth-grade exam correctly.[235]

One priority software automatically analyzes documents inside law firms, hospitals, banks, and other businesses. In response to this, Jeremy Howard, founder of Fast.ai, an independent lab based in San Francisco, said: "Each time we build new ways of doing something close to human level, it allows us to automate or augment human labor … This can make life easier for a lawyer or a paralegal. But it can also help with medicine."[236] That brings up the question: at this juncture, what exactly is the final goal of all this machine AI learning and speech effort? More to the point, how is it going to increase and improve, rather than eliminate, jobs for humans?

Transplants of Body Parts

The fight by the medical profession for freedom to essentially develop and carry out any new technology in the name of "medical advances" is now at a crucial point. Here are two examples. Head transplants are moving toward reality. Dr. Ren in China reported operations on nearly 1,000 mice in June

---

234 Steve Lohr, "Is There a Smarter Path to Artificial Intelligence? Some Experts Hope So," *New York Times*, June 20, 2018, https://www.nytimes.com/2018/06/20/technology/deep-learning-artificial-intelligence.html.

235 Cade Metz, "A Breakthrough for AI Technology: Passing an 8th-Grade Science Test," *Seattle Times*, September 4, 2019, https://www.nytimes.com/2019/09/04/technology/artificial-intelligence-aristo-passed-test.html.

236 Cade Metz, "Finally, a Machine That Can Finish Your Sentence," *New York Times*, November 18, 2018, https://www.nytimes.com/2018/11/18/technology/artificial-intelligence-language.html.

2015, just after he had successfully transplanted the head of one mouse to the body of another mouse.[237] His next move was on hominoids, primates, and then a human corpse in 2017. Ren was planning head transfers between live humans despite criticism about lack of ethics and oversight in Chinese experimental medicine.[238]

The Italian scientist Dr. Sergio Canavero also believed that head transplants were possible and intended to conduct the first surgery in 2017, with the assistance of Dr. Ren.[239] The first volunteer for a head transplant, Sergio Spiridonov, ultimately backed out (changed his mind, figuratively but not literally). Apart from questions about where a donor of a head would come from, head transfers raise unprecedented philosophical and ethical issues, such as what it would mean for a person's identity to have a new body, and how the mind and the body are deeply interconnected and a change in either component of the body-mind will have dramatic effects on the other. Arthur Caplan, a bioethicist, has dismissed Canavero's claims, writing, "Head transplants are fake news. Those that promote such claims and who would subject any human being to unproven cruel surgery merit not the headlines but only contempt and condemnation."[240]

A second example is 3-D printed replacements of body parts in humans, sort of an extension of cyborgized humans. In 2014, doctors in the Netherlands successfully implanted a plastic, 3-D-printed cranium replacement for a rare skull disorder.[241] It was reported in a 2015 news article that, for the first time, a partial skull and scalp transplant from a human donor was successful. After that operation was over, the man

---

237   Shirley S. Wang, "Surgery's Far Frontier: Head Transplants." *Wall Street Journal*, June 5, 2015, http://www.wsj.com/articles/surgerys-far-frontier-head-transplants-1433525830.

238   Didi Kirsten Tatlow, "Doctor's Plan for Full-Body Transplants Raises Doubts Even in Daring China," *New York Times*, June 11, 2016, http://www.nytimes.com/2016/06/12/world/asia/china-body-transplant.html?_r=0.

239   Ibid.

240   Albert Caplan, "Promise of world's first head transplant is truly fake news," *Chicago Tribune*, December 13, 2017.

241   Joel Lindsey, "Woman Has Entire Skull Replaced With 3D Printed Implant," Meddeviceonline, April 1, 2014, http://www.meddeviceonline.com/doc/woman-has-entire-skull-replaced-with-d-printed-implant-0001.

also got a new kidney and pancreas.²⁴² Currently, AI scientists researching brain augmentations to enhance intelligence are delving into the feasibility of using a 3-D-printed brain. In other words, just keep adding 3-D printed body parts ad infinitum until a truce is called and death with dignity is arranged for humankind as we know it.

Reproduction: Womb Transfer and Surrogate Babies

Other barriers are dropping for humanoids to be on par with humans, and facilitating reproduction is one of the crucial steps. A woman in Sweden, who had received a uterus in 2014 from a close family friend, is a medical first. She gave birth after receiving that womb transplant.²⁴³ This feat opens an experimental alternative for the thousands of women each year who are unable to have children because they lost their uterus to cancer or were born without one, and for other reasons.²⁴⁴ In a first for the U.S., a woman had a baby after a uterine transplant in 2017.²⁴⁵ In another first, this one for Latin America, a live birth in Brazil resulting from the implantation of the donor's egg into a transplanted uterus was reported in December 2018. The donor had died of a stroke. The mother and the baby were reported to be healthy nearly a year after the birth.²⁴⁶ And in 2020, the procedure was replicated in Philadelphia, leading to

---

242   Marilynne Marchione, "Texas Doctors Do First Skull, Scalp Transplant," MSN.com., June 4, 2015, http://www.msn.com/en-us/news/us/texas-doctors-do-first-skull-scalp-transplant/ar-BBkGvZP.

243   Maria Cheng, "Woman Gives Birth Via Womb Transplant," CTVnews, October 3, 2014, http://www.ctvnews.ca/health/in-world-first-woman-gives-birth-to-baby-after-womb-transplant-1.2038284.

244   Denise Grady, "Uterus Transplants May Soon Help Some Infertile Women in the U.S. Become Pregnant," *New York Times*, November 12, 2015, http://www.nytimes.com/2015/11/13/health/uterus-transplants-may-soon-help-some-infertile-women-in-the-us-become-pregnant.html.

245   Marilynn Marchione, "First baby from a uterus transplant in the US born in Dallas," Houston Chronicle, December 1, 2017, http://www.houstonchronicle.com/news/texas/article/First-baby-from-a-uterus-transplant-in-the-US-12398920.php.

246   Emily Baumgaertner, "From a deceased woman's transplanted uterus, a live birth," *Seattle Times*, December 5, 2018, https://www.seattletimes.com/nation-world/from-a-deceased-womans-transplanted-uterus-a-live-birth/.

another live birth.[247]

On the male side, penis transplants are in the works. Medical journals had reported two successful penis transplants as of the end of 2015.[248] The immediate hope is that injured men will regain urinary function, sensation, and eventually, the ability to have sex. It may sound bizarre, but womb and penis transplants between humans raises the possibility of similar transplants from humans to existing humanoids, and later, between advanced humanoids, which leads us to the possibility of surrogacy by gynoids.

One method of surrogacy, an arrangement often supported by a legal agreement, whereby a woman (the surrogate mother) agrees to become pregnant and give birth to a child for another person. In robotics, for a gynoid via transplants from humans. Another method is to impregnate the gynoid that has necessary organs, such as a uterus. There are two methods for impregnating potential surrogate mothers: naturally or artificially. The former method, referred to as *traditional surrogacy*, has been practiced the longest. Because the surrogate's own egg is used in the fertilization process, the resulting child is genetically related to the surrogate. "Before modern technology allowed for the creation of embryos outside the womb, it was the only way to conceive via a surrogate mother. Now that the means do exist to use the intended mother's or a donor's egg, traditional surrogacy is much rarer."[249]

The other procedure is *gestational surrogacy*, the procedure in which pregnancy is derived from transfer of an embryo created by in vitro fertilization in which a child is unrelated to the surrogate. This is the most common method in the United States because it is considered to be less complex legally. Naturally, there are sometimes conflicts about who gets custody of the child. This has been partially resolved by Iowa's Supreme Court conclusion, on February 16, 2018, that the woman who agrees to accept payment for having a baby is not legally the

---

247   Jacqueline Howard, "Second baby in the US born from transplanted uterus of deceased donor," CNN, January 9, 2020, https://www.cnn.com/2020/01/09/health/uterus-transplant-second-birth-us-bn/index.html.

248   Denise Grady, "Penis transplants in works to heal troops' hidden combat wounds," *New York Times*, December 7, 2015, http://www.nytimes.com/2015/12/07/health/penis-transplants-being-planned-to-heal-troops-hidden-wounds.html?_r=0.

249   ConceiveAbilities, "The Different Types of Surrogacy," August 10, 2018, https://www.conceiveabilities.com/about/blog/the-different-types-of-surrogacy.

child's parent and that surrogacy contracts for childbirth are enforceable.²⁵⁰

In another birth-related angle, "researchers are creating an artificial womb to improve care for extremely premature babies"—and remarkable animal testing suggests that the first-of-its-kind watery incubation so closely mimics mom that it just might work." ²⁵¹ Tiny lamb preemies receive treatment more like fetuses than newborns in the Biobag system. It takes a little imagination to conceive of Biobag experiments on advanced humanoid surrogate mothers. Yasuo Kuniyoshi, Director of the Intelligent Systems and Informatics Laboratory at the University of Tokyo, developed a robot fetus, a complete robot and computer simulation of a developing human body and nervous system. Olaf Groth and Mark Nitzberg reported that "the extremely precise model even floats in a liquid-filled 'womb' in which it wiggles and moves spontaneously. Through sensors and its neural network, the robotic fetus begins to learn about itself and its environment. So, if its limbs touch each other, he [Kuniyoshi] explains, the robot recognizes and learns about that physical relationship. 'That, in turn, changes its behavior and changes the output,' he [Kuniyoshi] says. 'As the baby learns it changes its behavior and changes its input, because the movement changes. If this continues on, we think we can call it a spontaneous development.'"²⁵²

Another achievement suitable for experimentation on humanoids is that a transgender woman was able to induce lactation.²⁵³ The thirty-year-old woman, who was born a male, told doctors she hoped to nurse the baby of her pregnant partner, and the doctors started giving her milk-inducing medications

---

250   David Pitt, "Court upholds surrogacy contracts as enforceable in Iowa," *Seattle Times*, February 16, 2018, https://www.seattletimes.com/nation-world/court-upholds-surrogacy-contracts-as-enforceable-in-iowa/.

251   Lauran Neergaard, "Hope for preemies as artificial womb helps tiny lambs grow," *Houston Chronicle*, April 25, 2017, http://www.houstonchronicle.com/news/nation-world/nation/article/Hope-for-preemies-as-artificial-womb-helps-tiny-11098824.php.

252   Olaf Groth and Mark Nitzberg, *Solomon's Code: Humanity in a World of Thinking Machines* (London: Pegasus Books, 2018), 214.

253   Ceylan Yeginsu, "Transgender Woman Breast-Feeds Baby after Hospital Induces Lactation," *New York Times*, February 15, 2018, https://www.nytimes.com/2018/02/15/health/transgender-woman-breast-feed.html.

(though one of those medications, an anti-nausea drug called domperidone, has not been approved by the FDA). The woman also took female hormones and stimulated her chest using a breast pump. "Within one month she was producing droplets of milk," and within three months she was producing eight ounces each day. She was able to feed the baby using only her breasts for six weeks, according to the study's authors.

Does it seem uncanny or impractical that advanced humanoids could be surrogate mothers for human or humanoid's babies? I think so, considering that the internet is full of surrogacy agreements and successes, achievements that were unthinkable not so long ago. Imagine an advanced humanoid, in around 2035 or 2040, as a partner in surrogacy. Ideological and legal questions that impinge on this case will have come to the forefront by then.

Embryo Transfer

Embryo transfer is an important developmental advance in humanoids. ET, as it is commonly called, refers to "a step in the process of assisted reproduction in which an embryo" (a new organism in the earliest stage of development) or "embryos are inserted into the uterus of a female with the intent to establish a pregnancy."[254] An easy way to visualize this technique is with cattle, in which case the purpose is rapid improvement in the herd's genetics.[255] The process is to collect somatic cells (any cell that makes up an organism, except for a reproductive cell) from a donor cow and manipulate them into an egg cytoplasm that results in a one-cell embryo. Those are then treated to produce a successfully developed embryo or multiple embryos. This procedure has been practiced for many years, as has splitting embryos into several parts in the maturation stage before

---

254   Wikipedia, "Embryo transfer," last updated July 21, 2020, https://en.wikipedia.org/wiki/Embryo_transfer#:~:text=Embryotransferreferstoa,intenttoestablishapregnancy.

255   James R. Simpson, "ETCALF: A Computer Program to Determine Bovine Embryo Transfer Costs" (Metric Version). Circular SW-094, Florida Cooperative Extension Service, November 1995; also, James R. Simpson and Shioya Yasuo, "Cattle Embryo Transfer Economics: The Koiwai Farm in Iwate Prefecture, Japan," 国際文化研究 *Kokusaibunka Kenkyuu (Intercultural Studies)* 3 (1999): 71-77. (In English).

transferring embryos into multiple animals. ET is not cloning, although it is a gradual predecessor of cloning. The process of embryo transfer is also required for other reproductive technologies, such as in vitro fertilization, in which an egg is combined with sperm outside the body.

Disputes over custody of embryos have been taking place for considerable time. Why wouldn't they be? After all, literally hundreds of thousands of embryos are left in limbo when couples divorce or simply disagree about how many children they want. The fights have now morphed into an epic legal battle over legal rights for embryos.[256] Most lawsuits are about frozen embryos left over from in vitro fertilization procedures. According to a *Washington Post* article, "Judges have often—but not always—ruled in favor of the person who does not want the embryos used. Sometimes they are destroyed, following the theory that no one should be forced to become a parent. The state of Arizona took the opposite approach. Under a first-in-the-nation law that went into effect July 1 [2018], custody of disputed embryos must be given to the party who intends to help them 'develop to birth.'"[257] What will happen to them after they are born? With conflicting rulings in various states, many predict the issue will ultimately be decided by the U.S. Supreme Court.

These legal cases presage other legal rights cases about embryos for enhancement of humans, humanoids, cloned humanoids, and embryos transferred to and between humans, and between humanoids. Make no mistake about it: human versus robot humanoid and/ or humanoid battles in America, over the time of conception, abortion, and women's rights to use their bodies as they wish, will play out as the symbiosis of humans and machines gains steam. After all, American's love-hate relationship with litigation and ethical issues ensures that the murky question of what constitutes a living organism

---

256 Tamar Lewin, "Anti-abortion groups join battles over ex-couples' frozen embryos," *New York Times*, January 19, 2016, http://www.nytimes.com/2016/01/20/us/anti-abortion-groups-join-battles-over-frozen-embryos.html.

257 Ariana Eun Cha, "Who gets the embryos? Whoever wants to make them into babies, new law says," *Washington Post*, July 17, 2018, https://www.washingtonpost.com/national/health-science/who-gets-the-embryos-whoever-wants-to-make-them-into-babies-new-law-says/2018/07/17/8476b840-7e0d-11e8-bb6b-c1cb691f1402_story.html?noredirect=on&utm_term=.2005811475df.

will be a hot button at some point. In brief, be prepared for legal battles that will arise over robot humanoid and humanoid innovations, development, and possession. The South Korean Robot Ethics Charter 2012 all but guarantees that scenario.

Part 2

CREATING HUMANOIDS AS LIVING ORGANISMS

By "life," I mean self-nourishment, growth, and decay.

—Aristotle, *On the Soul*

A frank conversation is needed about bans on existential risk technologies that realistically can lead to the demise of humankind. How about you? Is that what you would want to happen? "Not really," you say. But will the public demand controls, regulations, or bans on such extreme technologies within the next two decades? How about on humanoids reaching the level of being on par with humans knowing that further enhancements could result in radical humanoidization, the process of humanoids attaining such radically advanced capabilities that they would no longer be unambiguously human by our current standards?

SOME GENOMICS INNOVATIONS ARE TRULY UNNERVING

Genomics is a field within the discipline of genetics that focuses on determination and analysis of the structure of genomes (the complete set of DNA within a single cell of an organism, such as humans), and applications of DNA in current practice and innovations. One belief is that people will be able to monitor their health better if they have their DNA sequenced early in life. Goals to improve a human's health seem wonderful, beneficial, and laudatory, and they are, in the short and perhaps medium-term, time frames. The longer term is a different story, for it gives rise to greater possibility that the law of unintended consequences will come into play as the public is deluded into self-deception by only spotlighting the benefits side.

Unfortunately, there is also an unnerving advance in genomics that goes far beyond diseases and health concerns and that is potentially so perilous that some scientists seek a ban on it. It's a method of editing the human genome, developed in 2012.[258] At issue is CRISPR-Cas9, developed by Jennifer Doudna, leader and co-inventor of the technology, which is a revolutionary tool for editing DNA in plants, animals, and humans. The problem is that "the simplicity of the CRISPR-Cas9 enables any researcher with knowledge of molecular biology to modify genomes, making feasible many experiments that were previously difficult or impossible to conduct.[259] Tragically, the method will potentially open a floodgate to calamitous consequences. For example, researchers have realized that CRISPR-Cas9 might be used "not only to edit and repair genes of people living with diseases but also to edit embryos," with the intended and unintended consequences of changing the DNA of future generations. One negative repercussion would be "a so-called 'gene drive,' spreading engineered genetic changes through populations of wild animals but also altering the environment in unpredictable ways." [260]

Two Nobel laureates for gene technology, Dr. David Baltimore, president emeritus of the California Institute of Technology, and Dr. Paul Berg, Cahill Professor of Biological Chemistry Emeritus, argue that issues abound, despite alterations being quite precise. They write that if Crisper-cas9

---

258 Nicholas Wade, "Scientists Seek Ban on Method of Editing the Human Genome," *New York Times*, March 19, 2015, http://www.nytimes.com/2015/03/20/science/biologists-call-for-halt-to-gene-editing-technique-in-humans.html?_r=1.

259 Robert Sanders, "Scientists urge caution in using new CRISPR technology to treat human genetic disease," UC Berkeley, News Center, March 19, 2015, http://newscenter.berkeley.edu/2015/03/19/scientists-urge-caution-in-using-new-crispr-technology-to-treat-human-genetic-disease/.

260 Amy Dockser Marcus, "Communities Raise Their Voices on Genetic Engineering," *Wall Street Journal*, July 27, 2018, https://www.wsj.com/articles/communities-raise-their-voices-on-genetic-engineering-1532717088.

is used, "the decision to alter a germ-line cell may be valuable to offspring, but as norms change and the altered inheritance is carried into new genetic combinations, uncertain and possibly undesirable consequences may ensue."[261]

The two experts explain that medically necessary genome modifications can be achieved by creating embryos conventionally by using in vitro fertilization and implementation, followed by embryo selection. In contrast, the other, more unsettling kind of germ-line modification would involve attempts to modify inheritance for the purpose of enhancing an offspring's physical characteristics or intellectual capability. We call this *voluntary* modification in that there is no compelling medical need. Choosing to transmit voluntary changes to future generations involves a value judgement on the part of parents, a judgement that future generations might view differently....Baltimore and Berg continue, This can be seen as eugenics, thought by earlier generations to be desirable but now generally considered abhorrent. We also often do not know well enough the total range of consequences of a given gene alteration, potentially creating unexpected physiological alterations that would extend down through generations to come. For these reasons, and others, voluntary genome alteration might well be outlawed, at least in the present stage of knowledge. [262]

The larger critical question is whether there must be globally accepted moratoriums, regulations, or bans not only on CRISPR-Cas9 but also on technologies that exhibit existential risk. It did not take long to contravene the call for a voluntary moratorium on making changes to DNA that could be passed down to subsequent generations. British researcher Kathy Niakan, of the Francis Crick Institute in London, received permission on February 1, 2016, from the British regulatory agency that supervises reproductive biology, the Human Fertilization and Embryology Authority, to use the powerful genome editing technique CRISPR-Cas9 to alter human embryos.

Dr. Niakan's goal seemed benign: to improve knowledge about the basic biology of embryo development, rather than to discover a specific medical treatment. A news article noted

---

261 David Baltimore and Paul Berg, "Let's Hit 'Pause' Before Altering Humankind," *Wall Street Journal*, April 8, 2015.

262 Ibid.

that "Dr. Niakan, a developmental biologist, has no intention of implanting the altered embryos in a womb. According to a report in *Nature*, she will let the embryos expire when they are seven days old and have reached the blastocyst, or implantation, stage."[263] However, coincidentally, there was politically motivated news that the United States Congress "has forbidden the government to support research in which a human embryo is destroyed, although the ban does not apply to privately funded researchers."[264] On December 30, 2019, a court in China sentenced He Jiankui, the researcher who shocked the scientific community with his claim that he had used the CRISPR-Cas9 editing technique to create world's first genetically edited babies, to three years in prison.

Cloning

Our first step is to define what we are talking about. *Cloning*, in biology, is "the process of producing similar populations of genetically identical individuals that occurs in nature when organisms such as bacteria, insects or plants reproduce asexually."[265] *Cloning*, in biotechnology, refers to "the processes used to create copies of DNA fragments (molecular cloning), cells (cell cloning), or organisms (organism cloning)." *Organism cloning* (also known as reproductive cloning) is "the procedure of creating a new multicellular organism, genetically identical to another." This technology has been practiced in a very crude form—by grafting and other asexual propagation methods—since plants were first domesticated.

Asexual reproduction, a naturally occurring phenomenon in many species, "occurs when an organism makes more of itself without exchanging genetic information with another

---

263   Nicholas Wade, "British Researcher Gets Permission to Edit Genes of Human Embryos," *New York Times*, February 1, 2016, http://www.nytimes.com/2016/02/02/health/crispr-gene-editing-human-embryos-kathy-niakan-britain.html?_r=0.

264   Ibid.

265   Biology Online, s.v. "Cloning," n.d., https://www.biologyonline.com/dictionary/cloning.

organism through sex."²⁶⁶ Additionally, it is where fertilization or inter-gamete (*gametes* are an organism's reproductive cells) contact does not take place. It has been the foundation for plant improvements, at a higher level, for at least two millennia. *Therapeutic cloning*, the creation of embryonic stem cells for treatment of diseases such as diabetes is common, but not used on humans.

The concept of *reproductive cloning*, a technique that would involve making cloned humans, is quite controversial, with most scientific and governmental organizations opposing it based on safety rather than moral grounds. This type of cloning "generally uses the technique of 'somatic cell nuclear transfer' (SCNT) to create animals that are genetically identical. The process entails transfer of a nucleus from a donor adult cell (somatic cell) to an egg from which the nucleus has been removed, or to a cell from a blastocyst [a structure formed in the early development of mammals] from which the nucleus has been removed. If the egg begins to divide normally, it is transferred into the uterus of the surrogate mother. Such clones are not strictly identical since the somatic cells may contain mutations in their nuclear DNA."²⁶⁷

The first mammal cloned from an adult cell taken from her mother's udder in 1996 was a sheep named Dolly. It died in 2003. Cattle, goats, pigs, sheep, and many other species have all been successfully cloned. The FDA (Food and Drug Administration) approved meat and other products derived from these animals in 2006 for consumption in the United States, based on cloned animal products being "virtually indistinguishable"²⁶⁸ from non-cloned animal products.

Chinese scientists successfully cloned and generated two healthy cynomolgus macaque monkeys, using the SCNT procedure from adult cells, in January of 2018. The intention was to use these "genetically uniform monkeys as animal

---

266 Biology Dictionary, s.v. "Asexual Reproduction," updated January 28, 2020, https://biologydictionary.net/asexual-reproduction/#:~:text=Asexualreproduction-occurswhenan,withanotherorganismthroughsex.&text=Asexualreproductionispracticedby,plantsanimalsandfungi.

267 Joseph Wright, *Gene Control* (Scientific e-Resources, 2019), 171.

268 FDA Rules Beef From Cloned Animals Is Safe," *Beef*, December 29, 2006, https://www.beefmagazine.com/Food_From_Cloned_Animals_Safe.

models for basic research in primate biology and for studying human disease mechanisms and therapeutic treatments."[269] Chimpanzees now have the distinction of being our closest living relative in the animal kingdom, so this case has led to concerns about cloning in humans, since monkeys and humans are both primates. An international team of researchers has sequenced the genome of the bonobo for the first time, confirming that it also shares the same percentage—98.8—of its DNA with us as chimps do, which has heightened the concern..

The significant ethical concerns about harvesting organs directly from humans for transfer into other organisms such as pigs and cows, which then grow them for later transfer back to humans, is a form of xenotransplantation. Somewhat related to that is cloning extinct or endangered species. *Xenotransplantation* is the process of removing living cells, tissues, or organs from one species and implanting those elements in another species. Xenotransplantation is being researched and developed because there are many people waiting for organ transplants and not enough organs for all of them. On average, twenty people die each day waiting for an organ transplant.[270] Xenotransplantation might be the answer to the need for viable organs for transplantation. Another use of xenotransplantation would be to create human organs that could be used, in lieu of animals, for testing medications more accurately, since the efficacy of a drug regimen might be different in another species.

There are four different types of xenotransplantation. Solid organ xenotransplantation, the first type, is "a procedure in which a source animal organ such as kidney or liver is transplanted into a human." The second type, cell and tissue xenotransplantation, is "the transplantation of tissues and cells from a source animal without surgical connection of any animal blood vessels to the recipient's vessels." Extracorporeal perfusion, the third type, "occurs when human blood is circulated outside of the human body through an animal organ,

---

269   Zhen Liu et al., "Cloning of Macaque Monkeys by Somatic Cell Nuclear Transfer," *Cell* 172, no. 4 (February 8, 2018): 881-887.e7, https://doi.org/10.1016/j.cell.2018.01.020.

270   Health Resources and Services Administration, "Organ Donor Statistics," last reviewed June 2020, https://www.organdonor.gov/statistics-stories/statistics.html.

such as a liver or a kidney, or through a bioartificial organ produced by culturing animal cells on an artificial matrix." The final type, exposure to living animal-derived material, "is a procedure in which human body fluids, cells, tissues, or organs are removed from the body, come into contact with animal cells, tissues, or organs and are then placed back into a human patient." There are significant ethical concerns related to this last type of xenotransplantation.[271]

Scientists have begun to blur the line between human and animal through research on chimeras, single organisms made up of cells with distinct genotypes. "In animals, this means an individual derived from two or more zygotes, which can include possessing blood cells of different blood types, subtle variations in form (phenotype) and, if the zygotes were of differing sexes, then even the possession of both female and male sex organs."[272] These chimeras are in the embryonic stage only and being used to perfect the fourth type of xenotransplantation. The advantage in using chimeras in this process is that the resulting organ will be less likely to be rejected by the recipient's body. Issues about xenotransplantation and human cloning have been debated for quite some time, so it is not farfetched to imagine xenotransplantation of human's organs used in advanced humanoids, provided the vascular and pulmonary systems are also transplanted to support the organs. At that point, or around it, the creature could be cloned and replicated.

Why, it is reasonable to ask, should we humans at this time believe using xenotransplantation and chimeras will become an accepted practice? The short answer is because of technocrats' and government's overwhelming desires, power, abilities, and *freedom to* bring about such accomplishments.

Cloning Humans

Human cloning, generally referred to as artificial human cloning, is "the creation of a genetically (nearly) identical copy

---

271 World Health Organization, "GKT4 Xenotransplantation," n.d., https://www.who.int/transplantation/gkt/xenotransplantation/en/.

272 "Heredity–Traits–Genes," BK101 Knowledge Base, n.d., https://www.basic-knowledge101.com/categories/heredity.html.

of a human, which is the reproduction of human cells and tissues." Other approaches to cloning, such as the first hybrid human clone created in 1998 using DNA removed from a man's leg cell and a cow's egg, have led to continued ethical concerns about human cloning. This disquiet is present largely because cloning, with the possibility of creating humans, has been a significant issue in that the hybrid cell becomes a complete human embryo when it reaches fourteen days (which is when a normal embryo implants).[273] The 1998 hybrid human clone issue was bypassed by destroying the embryo at twelve days, along with the announcement that the aim was only to make advances in therapeutic cloning.

Seven years later, in 2005, Maryann Mott wrote that an experiment to genetically engineer mice to produce human sperm and eggs, and then to carry out in vitro fertilization to produce a child whose parents are a pair of mice, would raise concerns.[274] That procedure caused an uproar because it involves mixing human stem cells with embryonic animals to create new species. One simple definition of a species is a group of living organisms consisting of similar individuals capable of exchanging genes or interbreeding. Nevertheless, in 2008, relentless interest in the techniques led to cloning five mature human embryos using DNA from adult skin cells.[275] The stated aim was to provide viable embryonic stem cells (a stem cell is an undifferentiated cell of a multicellular organism that can give rise to indefinitely more cells of the same type). In this situation too, the legalities and ethical considerations were bypassed by destroying the embryos.

Many religious organizations are opposed to all forms of cloning based on the belief that life begins at conception. Currently, scientists do not intend to try to clone people. However, unintended consequences lurk in our rapidly changing world. The possibility of rogue scientists or ones

---

273 "Details of hybrid clone revealed," BBC News, June 18, 1999, http://news.bbc.co.uk/2/hi/science/nature/371378.stm.

274 Maryann Mott, "Animal-Human Hybrids Spark Controversy," *National Geographic News*, January 25, 2005, https://www.geneticsandsociety.org/article/animal-human-hybrids-spark-controversy.

275 Stemagen, "Cloned Embryo Created From Skin Cells," *ScienceDaily*, January 22, 2008, http://www.sciencedaily.com/releases/2008/01/080118092439.htm.

in countries with lax restrictions doing as they will without oversight calls for wider discussion and a public that is informed about the necessity of global laws and regulations governing this topic.

Freely permitting development of innovations related to human life brings concerns that at some point, attempts will be made to clone and transfer human embryos and chimeras into humanoids. Think back to Maryann Mott's concern that caused an uproar because it involved mixing human stem cells with embryonic animals to create new species.

Thinking, Emotions, and Brains for Being a Living Organism

Critical questions arise about attributes a humanoid must/should/could be endowed with so that a substantial portion of the population would opine, "I now consider this humanoid to essentially be on par with a human." Life in the biological sense is a characteristic that distinguishes physical entities from those with self-sustaining processes. In common parlance, life is the ability to breathe, grow, reproduce, and so forth, that people, animals, and plants all have before they die, but that objects do not have. In effect, for human beings, life is the state of being alive as a human being.

What, then, are the qualities necessary in a humanoid for a human to consider it alive? Must it have a network of vascular channels within a block of tissues that, much like blood vessels, can deliver nutrients to keep other parts of the synthesized human platform humanoid alive? Must it breathe? Can it be like a human cyborg that is dependent on an oxygen bottle for breathing? Are quality and source of speech prerequisites? Is it acceptable if a humanoid's computer-driven brain is programmed for realistic speech and abilities to make or take routine calls, as chatbots do to answer technical questions?

Is thinking and problem-solving at a level equal to that of humans a necessary condition for determining whether a humanoid is alive? Mathematicians, philosophers, physicists, and AI scientists wrestle with the question of whether a humanoid would be able to think and solve complex and theoretical problems equally as well as or better than humans. The debate continues because some believe that machines, at least in thinking and problem-solving skills, cannot or should not be developed to be on par with humans. Others disagree.

Let's move on to emotions. Humans have a wide range of emotional capacities and social skills that differentiate them from mammals. For example, some people are taciturn and rarely smile or display emotions, while others are very outgoing and effusive; or, depending on the circumstances, most people will be a little of both. As it happens, significant research is underway to unmask hidden emotions using psychology and facial-recognition software, as well as hardware such as cameras, to enable the capture of facial cues and gaze tracking, along with biosensors to measure physiological responses. Companies such as Affectiva, Emotient, Eyeris, and Sension are teaching computers to detect emotions such as surprise, happiness, boredom, or disgust in even the tiniest of facial expressions.

A few humanoid robots and humanoids are already programmed with image recognition to recognize how a person feels and to respond accordingly. It is my opinion that when it comes to humanoids being on par with humans, powerful computational techniques and faster and cheaper computers guarantee that advanced humanoids will be programmed with even more subtle, emotional, humanlike reactions will have the ability to update themselves, unless controls are placed on them by their designers.

Now let's look at a brief introduction to the makeup of brains. Some researchers argue that a full understanding of the brain is further ahead than is commonly recognized and that only a few decades will be required before bionic brains are programmed to think and reason. But it looks as if that is already happening. Programmed computers are beating top international chess players. Google's artificial-intelligence program AlphaGo's win against the top Go player in May 2017 is evidence of how computers could surpass humans in complex tasks; it's also a victory with "a glimpse of the promise of new technologies that mimic the way the brain functions."[276]

---

276   Paul Mozur, "In a win for artificial intelligence, Google computer program defeats Chinese Go master," *Seattle Times*, May 23, 2017, http://www.seattletimes.com/business/in-a-win-for-artificial-intelligence-google-computer-program-defeats-chinese-go-master/.

Mathematician Alan Turing (1912–1954) is widely considered a founder of artificial intelligence. Some of his followers argue that machines we create are not, and cannot be, one-for-one replacements for humans. The argument over whether machines can be more intelligent than humans has been going on for a long time. The Turing test, originally called the imitation game by Alan Turing in 1950 is a test of a machine's ability to exhibit intelligent behaviour equivalent to, or indistinguishable from, that of a human. Cowan argues in his chapter "Why the Turing Game Doesn't Matter" in his 2013 book, *Average Is Over: Powering America Beyond the Age of the Great Stagnation* that the Turing test really doesn't test human intelligence, because great weight is given, among scientists, to whether intelligent machines can pass a Turing test. The notion is that if the machine can consistently "pass" as a human on this benchmark, the machine has succeeded at the test and is truly intelligent.

Cowan points out that Alan Turing's core message, inability to imitate, does not rule out intelligence; rather, if a machine can imitate a man, the machine must be intelligent. In effect, Turing's only focus was on intelligence and machines. He further argues, "We wish genius machines to serve our practical ends, but we don't want to turn over to them the spheres of life that structure our narratives, drive our emotions, define what our lives are all about, and help us separate right from wrong. We're determined to 'keep them in their place'. ... For better or worse, we will remain—at the margin—rather desperately in need of help from the genius machines."[277]

S. Barry Cooper, a mathematician at Leeds and the foremost scholar on Turing's work, holds there is "some scientific basis for the view that humans are doing something that a machine isn't doing—and that we don't even want our machines to do.....Our machines are not like us. We could make them like us if we want ... by putting them in mechanical bodies and raising them like children."[278] He notes that the future of

---

277 Tylar Cowan, *Average is Over: Powering America Beyond the Age of the Great Stagnation*. New York: E.P. Dutton, 2013.

278 Christopher Mims, "Why Humans Needn't Fear the Machines All Around Us," *Wall Street Journal*, November 30, 2014, http://www.wsj.com/articles/why-we-neednt-fear-the-machines-1417394021.

technology isn't about replacing humans with machines—"it's about figuring out the most productive way for the two to collaborate. In a real and inescapable way, our machines need us just as much as we need them."[279]

SHOULD HUMANOIDS BE GIVEN FREE WILL?

Free will is defined in various ways. One is the ability of agents to make choices free from certain kinds of constraints. Another is the power to make one's own decisions with or without control by God or fate. The meaning of life—its significance, origin, purpose, ultimate fate, and point at which life begins—is an important ethical concept and question. The philosophically daunting issues concerning free will and determinism range from whether humans even have free will and, particularly within the religious community, whether humans' actions are determined by an omnipotent divinity. However, as the Enlightenment dawned, it was no longer possible to take free will as a matter of faith. At that point, philosophers tried to argue that as long as we might have done differently in a similar, even if not identical, situation, we have free will.

The objective here is not to argue about what constitutes free will. Rather, and strangely enough, the free will conundrum applies directly and indirectly to those scientists and others involved in superintelligence and humanoid development, to the extent that they believe they have created any and all innovations.

The fact is, we cannot escape the free will issue related to development of humanoids given popularization of androids and cyborgs in science fiction, movies, and TV series. In the case of humanoids, there are important issues about rights to use to use their bodies and lives in the way they wants to, rather than having those issues be decided by society. That is where debates about AI come in. John McCarthy (1927–2011) is the person who coined the term *artificial intelligence*. His pioneering work, in which he defined AI as "the science and engineering of making intelligent machines," helped spawn a whole AI industry. McCarthy tackled problems in AI that

---
[279] Ibid.

are now crucial to giving free will to humanoids when he wondered whether it was "legitimate to ascribe certain beliefs, knowledge, free will, intentions, consciousness, abilities, or want to a machine or computer program."[280]

Free will to McCarthy was not an all-or-nothing thing. His view was that some agents have greater free will, or free will of different kinds, than others do. His objective was primarily technological in that he focused on what can make robots more useful. However, as he explained in his 1999 paper, he also studied aspects of free will in which he distinguished between having choices and being conscious of those choices. He considered that both are important, even for robots, as he wrote, "consciousness of free will requires more structure in the agent than just having choices and is important for robots. Consciousness of free will is therefore not just an epiphenomenon of structure serving other purposes.... Human free will is a product of evolution and contributes to the success of the human animal. Useful robots will also require free will of a similar kind, and we will have to design it into them."[281]

McCarthy's arguments about the multitude of ethical questions concerning the desirability of humans to extend free will to humanoids are serious and realistic. His views on robot consciousness are echoed by David Levy in his extensively researched book *Robots Unlimited: Life in a Virtual Age*, in which he declares: "There seems to be little reason to doubt that robots can and will have beliefs, even if they are only simulated beliefs. What is important here is that, if they convince us that they have beliefs, by what they say to us and how they act, then, in accordance with Turing's doctrine, we should accept them as having beliefs....From a position in which we believe robots to have beliefs, it is only a small step for us to accept that robots can be religious."[282]

Suppose humans were to program humanoids with the ability to think and reason, and as well with free will.

---

[280] Jack Schofield, "John McCarthy obituary," *Guardian*, October 25, 2011, http://www.theguardian.com/technology/2011/oct/25/john-mccarthy.

[281] John McCarthy, "Free Will—Even for Robots," Abstract, Computer Science Department, Stanford University, 1999, http://www-formal.stanford.edu/jmc/free-will/freewill.html.

[282] David Levy, *Robots Unlimited: Life in a Virtual Age* (Wellesley, Massachusetts: A K Peters, Ltd., 2006), 391.

Could those humanoids be expected to embrace the virtues of happiness that we cherish, remembering that, for Aristotle, happiness is the ultimate end and purpose of human existence? Now picture those humanoids that choose to reprogram themselves to the extent that they would no longer be unambiguously human by our current standards, which implies they would have the ability to make choices that could adversely affect humans. Would they be likely to choose happiness for themselves and the remaining humans as the end purpose of their existence?

HUMANOIDS AS A NEW SPECIES

A common definition of species is "a group of individuals that actually or potentially interbreed in nature and is the biggest gene pool possible under natural conditions." However:
> that definition of a species might seem cut and dried, but it is not—in nature, there are lots of places where it is difficult to apply this definition. ... For example, many bacteria reproduce mainly asexually. ... The definition of a species as a group of interbreeding individuals cannot be easily applied to organisms that reproduce only or mainly asexually. ... If two lineages of oak look quite different, but occasionally form hybrids with each other, should we count them as different species? There are lots of other places where the boundary of a species is blurred. It's not so surprising that these blurry places exist—after all, the idea of a species is something that we humans invented for our own convenience!"[283]

Humanoids can become a recognizable new species in a number of ways. One is by gradually developing, through augmentations and enhancements, into an advanced and complex stage at which humanoids are considered living creatures, and then to the extent that humans define them as a separate species. A second way is by constructing an identical copy or copies of itself via self-replication or cloning between humanoids or with humans. In brief, with cloning and other

---

283 "Defining a species," Evolution Berkeley, n.d., https://evolution.berkeley.edu/evo101/VADefiningSpecies.shtml.

techniques, if radically enhanced humanoids were deemed to be living organisms on a par with humans, especially if endowed with free will, they could choose to break apart from humankind and evolve into a new species.

GLOSSARY

(Last revision January 22, 2021)

Thanks is given to the multitude of sources I used that include dictionaries, encyclopedias, technical books, Wikipedia, etc., and for references and footnotes in the same. Many of these definitions are open to debate because the technical aspects surrounding them are changing so rapidly. In any event, the definitions are to help a wide swath of the public to understand the topic concerned, not as a definitive collection for technologists.

Actroid – a humanoid robot that looks very much like a real human.

Advanced humanoids – the term is applied to those in two stages. The initial one, in 2030-2035, in which they are moderately augmented and mix with humans, and 2035-2040, in which early advanced ones are generally accepted by humans, and the later part, when many are on par with humans.

Android – a humanoid robot or humanoid created as a synthetic organism designed to look and act as much as possible like a real person. The term popularly used for both males and female robots. The word has roots in androgynous, having the characteristics or nature of both male or female. Technically, an android is the male form. Gynoid is the female form. Droid is an abridgement of android.

Animaloid – a robot created as a synthetic organism designed to look and act as much as possible like a real animal.

Anthropocentric – regarding humankind as the central or most important element of existence, especially as opposed to God or animals. Considering humans and their existence as the most important and central fact about life.

Artificial Insemination – Injection of semen into the vagina or uterus other than by sexual intercourse.

Artificial Intelligence – (Definition 1) machinery with the ability to reason and solve problems. It also refers to the branch of computer science that includes the study and design of intelligence agents, the melding of humans, robots and machines, and aims to create intelligence of machines and robots.

Artificial Intelligence – (Definition 2. Oxford Dictionary) the theory and development of computer systems able to perform tasks normally requiring human intelligence, such as visual perception, speech recognition, decision-making, and translation between languages.

Automaton – a self-operated machine. A moving mechanical device made in imitation of a human being. A machine that operates on its own without the need for human control, or a person who acts like a machine.

Autonomous – capable of acting independently, without outside control. Autonomous machines can determine what actions to take without human direction.

Avatar – an image that represents you on the screen in an online game or chatroom or a person melding their mind and movements with a robot surrogate, or avatar.

Basic Humanoids – those in the basic stage of development. Also, the term used for androids because they have appearance, mobility, vision, and ability to defend themselves and not harm others.

Bio-android – used interchangeably with the term android.

Biological Engineering, also termed biotechnological engineering or bioengineering – an engineering discipline that combines methods and concepts in biology with those of computer sciences, mathematics, physics and chemistry to solve real world life sciences problems by using the engineer's knowledge of complex artificial systems.

Glossary

Bionics – application of biological method and systems found in nature to the study and design of engineering systems and modern technology. It also has come to include merging organism and machine, also referred to as a cybernetic organism, bionic person, or cyborg.

Bionic brains – artificial brain.

Biorobot – biologically inspired robot.

Biorobotics – the field focused on the construction of biologically inspired or biometric robots.

Bot – a device or piece of software that can execute commands, reply to messages, or perform routine tasks, as online searches, either automatically or with minimal human intervention.

Biotechnology (biotech) – use of living systems and organisms to develop or make useful products. Additionally, it is any technological application that uses biological systems, living organisms, or derivatives thereof, to make or modify products or processes for specific use.

Blastocyst – a structure formed in early development of mammals.

Blastocyte – an undifferentiated embryonic cell.

Brain Emulation (see Whole brain emulation).

Chatbot – computer program that can simulate human conversation. Examples include personal assistants such as Siri and chatbots that answer customers' questions on company websites.

Chimera – single organism composed of cells with distinct genotypes. In animals, this means an individual derived from two or more zygotes, which can include possessing blood cells of different blood types, subtle variations in form (phenotype) and, if the zygotes were of differing sexes, then even the possession of both female and male sex organs.

Cisgenic – the resulting organism when genetic material from the same species, or a species that can naturally breed with the host, is used.

Cloning – the process of producing similar populations of genetically identical individuals that occurs in nature when organisms such as bacteria, insects, plants, or animals reproduce asexually.

Cobot – or co-robot (from collaborative robot) is a robot intended to physically interact with humans in a shared workspace. The term is also used for humanoids that work hand in hand with humans.

Cybernetic human – incorrect usage as a synonym for a humanoid.

Cybernetic organism – commonly known a cyborg, it is essentially the transfer of technology between engineered forms and life forms.

Cyborg – an organism that has enhanced abilities due to augmentations and enhancements, particularly mechanical parts. A stricter definition is enhancing normal capabilities. The general use is for physical attachments within or on humans. For example, a human fitted with prosthetic leg, mechanical parts in knee surgery, pace makers, and hearing aids. In science fiction, a creature that is part human, part machine.

Cyborgization – endowment within or on humans of a metaphysical or physical attachment.

Deep learning – process in which multilayered neural networks are exposed to vast amounts of data. On their own, the networks learn to analyze the data and draw conclusions.

DNA – a molecule composed of two chains that coil around each other to form a double helix carrying the genetic instructions used in the growth, development, functioning, and reproduction of all known living organisms and many viruses.

Droid – abridgement of android.

Embryo transfer – a step in the process of assisted reproduction in which embryos are placed into the uterus of a female with the intent to establish a pregnancy.

GLOSSARY

Enhanced humans – see Human enhancement.

Existential Risk – a risk that cannot be undone that poses permanent large negative consequences to humanity.

Exoskeleton – an outer framework worn by a person that may be powered to assist the wearer in boosting strength and endurance. A rigid external covering for the body in some invertebrate animals, especially arthropods.

Expert systems – computers that store vast amounts of information about a specific field, such as business medicine. Expert systems are also programmed with detailed rules about how to process the data.

Extropianism – a philosophy of or belief in an evolving framework of values and standards for continually improving the human condition.

Facultative – In biology, it means organisms that can live with another organism, but do not have to. In contrast, some organisms are obligate, meaning they depend on another for survival.

Gamete – a haploid (term used when a cell has half the usual number of chromosomes) cell that fuses with another haploid cell during fertilization (conception) in organisms that sexually reproduce.

Genetic engineering (GE) – also called gene modification, it is the direct manipulation of an organism using biotechnology methods to alter the genetic makeup of an organism.

Genoid – the female form of an android. Technically, android is the male form.

Genetically modified organism (GMO) – an organism generated through genetic engineering.

Genomics – the field within the discipline of genetics that focuses on determination and analysis of the structure of genomes (the complete set of DNA within a single cell of an organism such as humans). It also includes efforts to determine the entire DNA sequence of organisms and to map them.

Hominoids – a primate of a group that includes humans, their fossil ancestors, and the anthropoid apes.

Homo sapiens – the kind of species of human being that exists now.

Humans – member of the species Homo sapiens; a human being, especially a person as distinguished from an animal or (in science fiction) an alien. Humans are multicellular organisms.

Human enhancement – attempts to temporarily or permanently overcome current limitations of the human body by the use of technological means to select or alter human characteristics and capacities whether or not the alterations result in bringing about characteristics and capacities that lie beyond the existing human range.

Humanoid – a robot based on the general structure of a human. Also, the term generally used in place of android or humanoid robot in an effort to humanize the mechanical being and make it more acceptable, and in many cases, loveable. In science fiction, the term humanoid is most commonly used to refer to alien beings with a body plan that is generally like that of a human, including upright stance and bipedalism.

Humanoid robot – a mechanical or artificial device more robot than humanoid in the basic stages leading to the advanced humanoid stage.

Humanoids, Advanced – see Advanced humanoids.

Humanoids, Basic – see Basic humanoids.

Humanoids, radically enhanced – see Radically Enhanced Humanoids

Humanity+ – Humanity Plus is an international organization derived from rebranding to project a more humane image from

The World Transhumanist Association (WTA), which advocates the ethical use of emerging technologies to enhance human capacities.

Humanoidization – the process of developing robots through symbiosis with humans to the extent advanced humanoids are common and on par with humans.

Humanoidization, Radical – the process of humanoids attaining such radically advanced capacities that they would no longer be unambiguously human by our current standards.

In vitro fertilization – the process of fertilization by extracting eggs, retrieving a sperm sample, and then manually combining an egg and sperm in a laboratory dish. The embryo(s) is then transferred to the uterus.

Intelligence – Following are uses of this term connected with superintelligence[284]

> Friendly AI – superintelligence whose goals are aligned with ours. Alternative: friendly artificial intelligence (also friendly AI or FAI) is a hypothetical artificial general intelligence (AGI) that would have a positive (benign) effect on humanity. It is a part of the ethics of artificial intelligence and is closely related to machine ethics.
>
> Whereas machine ethics is concerned with how an artificially intelligent agent should behave, friendly artificial intelligence research is focused on how to practically bring about this behaviour and ensuring it is adequately constrained.
>
> Artificial General Intelligence (AGI) – the ability to accomplish any cognitive task at least as well as humans. Alternative: the hypothetical intelligence of a machine that has the capacity to understand or learn any intellectual task that a human being can.

---

284   Taken from Max Tegmark, *Life 3.0: Being Human in the Time of Artificial Intelligence* (New York: Knopf, 2017), 39, and Stephanie Sammartino McPherson, *Artificial Intelligence: Building Smarter Machines* (Minneapolis: Twenty-first Century Books, 2018), 97.

General Intelligence – ability to accomplish virtually any goal, including learning.

Intelligence – the ability to accomplish complex goals.

Intelligence explosion – recursive self-improvement rapidly leading to superintelligence.

Narrow Intelligence – the ability to accomplish a narrow set of goals, e.g., play chess or drive a car.

Singularity – intelligence explosion.

Strong AI – AGI.

Superintelligence – general intelligence far beyond human level.

Synthetic intelligence – an alternative term for artificial intelligence which emphasizes that the intelligence of machines need not be an imitation or in any way artificial; it can be a genuine form of intelligence.

Universal intelligence – ability to acquire general intelligence given access to data and resources.

Knockout organism – the result when genetic material is removed from the target organism. Knockouts are used to study gene function, usually by investigating the effect of gene loss.

Law of Accelerating Returns – theory by Ray Kurzweil that electronic development, such as improvement in the speed, memory, and power of computers, proceeds at a rate that is continuously doubling.

Life – a principle or force that is considered to underlie the distinctive quality of animate beings, a process that can retain its complexity and replicate.

Machine ethics – the concern with how an artificially intelligent agent should behave. See also: Friendly AI.

GLOSSARY

Mind control – originally known as brain washing or thought control, it is increasingly being researched a way to manipulate or subvert an individual's thinking, behavior, emotions or decision by outside sources. The most prevalent and recognized form of mind control is through deep brain stimulation by drilling holes in the brain and inserting powerful electrodes to treat a wide range of disorders. Another form termed mind control is a technique that allows humans to interact with their surroundings through so-called avatars.

Nanobots – tiny robots made of DNA that can walk, pivot, and even work with microscopic forklifts.

Nanotechnology – the production and use of machines that are only slightly larger than atoms and molecules. Some researchers believe AI will combine with nanotechnology to transform medicine and other disciplines.

Neural nets – computer systems that loosely mimic the workings of the human brain.

Neuroscience – science of the nervous system, traditionally seen as a branch of biology, which now covers a multitude of other fields and is an integral part of Artificial Intelligence.

Nucleotide – the basic structural unit of nucleic acids such as DNA

Obligate – In biology, some organisms are obligate, meaning they depend on another for survival. Others are termed facultative, meaning they can live with another organism, but do not have to.

Organism – a living thing, it is the smallest contiguous unit of life.

Pharmacogenomics – studies of how a person's genetic makeup affects his or her body's response to drugs.

Pharming – a combination of farming and pharmaceutical refers to the use of genetic engineering to insert genes into host animals or plants that would otherwise not express those genes, thus creating a genetically modified organism (GMO).

Plutocracy – formally, government by the wealthy. In more general parlance, it is any form of government in which the wealthy exercise the preponderance of power, whether it is direct or indirect.

Plutocrat – a person who is powerful because of their wealth.

Posthuman – possible future beings whose basic capacities so radically exceed those of present humans, as to be no longer unambiguously human by our current standards.

Posthuman condition – state following enhancement so extreme some individuals would no longer be humans by our current standards and could choose to overcome humans, leaving at most, a tiny fraction of all humans to enjoy the benefits of posthumanity.

Posthumanoid condition – state following radical humanoidization in which some humanoids would no longer be on par with humans by our current standards and choose to overcome humans leaving at most a tiny fraction of all humans to enjoy the benefits of humanity as they know it and/or a posthuman condition.

Posthumanism – a term not used by transhumanists, but sometimes incorrectly used as a synonym for transhumanism.

Posthumanity – the result from creation of superintelligence that, while it may not entail the extinction of literally all intelligent life, it nevertheless constitutes an existential risk because the future that would result would be one in which a great part of humanity's potential had been permanently destroyed.

Posthumanoid – possible future beings whose basic capacities so radically exceed those of advanced humanoids, as to be no longer unambiguously humanoids by our current standards.

Post-singularity – the point beyond which there is no distinction between human and machine.

Radically enhanced humans – those with advanced capacities that they would no longer be unambiguously human by our current standards.

GLOSSARY

Radically enhanced humanoids – those in the radically advanced stage of development or beyond with such radically advanced capacities that they exceed those of present humans. At that point, advanced humanoids would no longer be unambiguously human by our current standards.

Recombinant DNA (rDNA) – molecules that are DNA sequences derived by molecular cloning methods that create new DNA sequences that would not otherwise be found in the genome.

Robot – a mechanical or artificial device primarily guided by a computer program or some electronic method.

Robot, Advanced – robots accepted by humans for their workplace and society because they have appearance, mobility, and other useful attributes.

Robot, Autonomous – stand-alone system, complete with its own computer termed the controller. The most advanced example is the smart robot.

Robots, Basic – the term I use for robots that have the most elemental attributes as machines. Synonym for humanoid robot.

Robots, Fourth Generation – in the research and development phase that includes artificial intelligence, self-replication, self-assembly, and nanoscale size.

Robot, Smart – stand-alone robot, which has a built-in AI system that can learn from its environment and experience to build on those capabilities and knowledge.

Robots, Swarm – work in fleets with all members under the supervision of a single controller.

Robots, Telepresence – simulate the experience and some of the capabilities of being present. Examples are home monitoring, smart houses, remote business consultations, and many other possibilities.

Roboethics – the area of study concerned with what rules should be created for robots to ensure their ethical behavior and how to design ethical robots.

Roboethical standards – ones related to rights of robots and interactions with humans.

Robotics – the branch of technology that deals with the design, construction, operation, and application of robots.

Robotoid – an artificial lifeform created through processes that are totally different from cloning or synthetics. Also, a small robot, child sized or smaller.

Robot fetishism – the fetishistic attraction to humanoid robots. It also refers to people acting like robots or dressed in robot costumes.

Sapience – the quality of being wise, or wisdom.

Sapient – a human of the species.

Sentient – capable of thinking and feeling; being aware of one's existence.

Sexbots – humanoids developed for personal sexual satisfaction or as sex workers in commercial establishments.

Singularity – also known as technological singularity. See technological singularity.

Singularitarianism – a movement defined by the belief that a technological singularity—the creation of superintelligence—will likely happen in the near future, and that deliberate action ought to be taken to ensure that the singularity benefits humans.

Species – group of individuals that actually or potentially interbreed in nature and is the biggest gene pool possible under natural conditions.

Social robotics – study of how robots, humanoids and humans learn to relate to each other.

Glossary

Somatic cells – any cell of a living organism other than the reproductive cells.

Sophont – An intelligent being; a being with a base reasoning capacity roughly equivalent to or greater than that of a human being. The word does not apply to machines unless they have true artificial intelligence, rather than mere processing capacity.

Superintelligence – Any intellect that greatly exceeds the cognitive performance of humans in virtually all domains of interest.

Symbiont – an organism living in a state of symbiosis or in a symbiotic relationship.

Symbiote – synonym for symbiont.

Symbiosis – commonly defined as a relationship between people, companies, etc. that is to the advantage of both. A generally accepted definition in biology is the living together of unlike organisms, which has been broadened to cover all species.

Synthetic Biology – an interdisciplinary branch of biology that combines disciplines such as biotechnology, evolutionary biology, molecular biology, systems biology, biophysics, computer engineering, and genetic engineering.

Technium – a network of different technologies all working together to support each other that operates as if it is a sentient being.

Technocentrism – a term that denotes a value system centered on technology and its ability to control and protect the environment.

Technocracy – a social or political system in which people with scientific knowledge have a lot of power; a sense of being governed primarily by technical experts; a meritocracy composed of those with power derived from scientific knowledge.

Technocrat – a term for a member of a powerful technical elite or someone who advocates the supremacy of technical experts; an expert in science, engineering, etc. who has a lot of power in politics and/or industry; those that have inordinate power and control through technology over society.

Technium – Kevin Kelly's term for a network of different supporting technologies all working together to support each other that operates as if it is a sentient being.

Technological singularity – the theoretical emergence of superintelligence through technological means. A hypothetical moment when artificial intelligence, human biological enhancement, or brain-computer interfaces will have progressed to the point of a greater-than-human intelligence that will radically change civilization, and perhaps even human nature.

Technology – the collection of tools, including machinery, modifications, arrangements and procedures used by humans, defined by Kevin Kelly as anything the mind produces.

Technomancy – imaginary or fictional category of magical abilities that affect technology. Also, magical powers gained through the use of technology.

Techosexuals – devotees of robot fetishism.

Transgenic – the process when genetic material from an unrelated organism is added to the host organism.

Transhumanism from Version 2.1 of *The Transhumanist FAQ* – viewed as an extension of humanism, from which it is partially derived. Transhumanist's emphasize that while humans and individuals' matter, that by promoting rational thinking and rational means the human organism can be improved. They argue that technological means can be used beyond traditional humanistic methods to eventually enable humans to move beyond what some would think of as "human."

Transhumanism, alternative – An ideology and movement that affirms the possibility and desirability of improving the human condition by overcoming fundamental human limitations that seek to guide us to a posthumanity condition.

Transhumanist – someone who advocates transhumanism.

GLOSSARY

Transhuman – intermediary form between the human and posthuman.

Triumvirate – In the past, referred to a group of three men responsible for public administration or civil authority. In the present, a triumvirate refers to a group of people representing three instruments of power: the technocracy, the plutocracy, and our national level government that while not a cabal, and not coordinated, collectively have the power to decide on superintelligence and radical humanoidization that can cause serious disruption to humankind, as we know it.

Turing test – a test performed to determine a machine's ability to exhibit intelligent behavior. The basic concept behind the test is that if a human judge is engaged in a natural language conversation with a computer where he cannot reliably distinguish machine from human, the machine passes the test.

Virtual reality – a simulated environment that your senses perceive as real.

Whole brain emulation (WBE) – Mind upload or brain upload (sometimes called "mind copying" or "mind transfer") is the hypothetical process of scanning the mental state (including long-term memory and "self") of a particular brain substrate and copying it to a computer. The computer could then run a simulation model of the brain's information processing. Then it responds in essentially the same way as the original brain (i.e., indistinguishable from the brain for all relevant purposes) and experiences having a conscious mind.

World Transhumanist Association (WTA) – founded in 1998 and focused on recognition of transhumanism as a legitimate subject of scientific inquiry and public policy. The WTA changed its name to "Humanity+" in 2008 as part of a rebranding to project a more favorable humane image. It launched H+ Magazine and in 2010, the magazine transitioned into a web-only publication.

Xenotransplantation – the process of removing living cells, tissues, or organs from one species and implanting the living cells, tissues, or organs in another species.

Index

A

Adler, Jonathan, 96
Artificial Intelligence, *see also* AI
   Aim, 51, 53, 61-62, 166
   agent, anthropomorphic sense, 64
   consequences, unintended, 72, 90, 160, 166
   International Policy Guidelines, 66, 89
   intelligence agents, 61-62, 65
   intelligent life, 2, 35, 40, 41, 60
   intelligent machines, 1, 18, 41, 54, 69-70, 134, 169-170
   prognostications, 11-13, 15, 69
   artificial general intelligence (AGI), 98, 181
   rogues, 2, 58, 166
Allen Institute for Artificial Intelligence, 90, 150-151. *See also* Aristo,
Allen Institute, Project Alexandra, 150
Amazon, 146
   Alexa 149, 150
   Echo, 150
   voice controlled virtual assistants, 150
America, 1, 3
   dream, 85
   exceptional, 77
   statuary power, 99
   wellbeing, 2, 85, 115, 137
American Humanist Association (AHA), 36
Android, 7, 8, 9, 10, 170
Animals, 20, 27, 29, 64, 129, 144, 146, 157, 160, 163-167
   animaloid, 18, 20-22
   dogs, 19-20
   rights, 104, 164
   pets, 19
   Paro (s), 8, 19, 29
Anthropomorphic sense, 64
Apple's Seri, 149
Aristotle, 137, 151, 172
Armageddon, 4, 74
Arrison, Sonia 54-57
Artificial neural network (ANN), 46
Asilomar, 61-63, 70, 99
Asimov, Isaac, 9, 23
Association for the Advancement of Artificial Intelligence (AAAI), 60
Atkinson, Dr., 111
Automation, 8, 71, 108-110

B

Baker, Gerald, 136
Baltimore Dr., David, 160-161
Benefit the dead, three-D, 148
Biobag experiments, 155
Biology, 160-162, 164, 176,
Bionics, 44, 46
Bioprinted tissues, implanted into people, 148
Biorobotics, 144
Biotechnology, 40, 162
Blethen, Frank, 83-84
Blue-collar, workers, ii

Body parts, 13, 56, 139, 147-148, 151-153
Bogle, John, 79
Bostrom, Nick, 33, 70
Brain (s), 13, 31-32, 38, 48-52, 130, 140, 168
Brain computer, 51, 72
Brooks, David, 137
Brown, Dan, 32
Bryan, William Jennings, 75, 88
Brynjolfsson, Erik, 65
Buddy, home robot 145
Proof, 91, 100-101, 125
Burke, Edmund, 134

C

Cabal, ii, 76
Call to action, 136
Cambridge Centre for the Study of Existential Risk, 72
Cameron, James, 92, 104
Capitalism, 78, 83, 105
Cartagena Protocol Biosafety, 94
Catch-22, 88
Center for Governance of AI, 68
Chatbot (s), 8, 21, 29, 61, 150, 167
Chimeras, 165
Jiankui, He, 162
Choice(s), 65, 89, 132-133, 135, 138
Chomsky, Noam, 77
Christiansen, Sonja Boehmer, 91
Citizens United ruling, 78
Civility, 82, 129
Clayton, Cornell, 82
Cloning, 13, 43, 140, 157, 162-166, 172
Cobots, 8, 143
Cognitive systems, 60
Companies, 8, 21, 49, 69, 81, 89, 90, 106, 120-121, 143, 146, 149-150, 168

Computer vision, 145-146
Congress, 4, 66, 75, 78, 81, 82, 162
Congressional Research Service, 14
Cooper, Ann Julia, 138
Corporate cronyism, 80
Cost-benefit analysis (CBA), 102-103
Cousteau, Jacques and Jacques-Yves, 59
Creative destruction, 105
CRISPR-Cas9, 160-162
Cui, Ang, 24
Cyborg, 48, 144, 167

D

DARPA, 49
de Bary, Heinrich Anton, 142
de Grey, Aubrey, 55
Deep learning, 146, 150
Right to control, 4, 133-134
Doherty, Brian, 118

E

Economics, 20, 21
   economy, 71, 83, 87, 103, 105-107, 110-112, 114-116, 118-120
   debt, 83, 92, 107-108, 116
   three sectors: industry, services, and agriculture, 107
Edgar, Richard, 82
Elite, 11, 31, 76, 86-87
Elites, 4, 44, 85, 87, 135
Embryo, 13, 139, 154, 156-157, 161-162
Embryos, 167, 178, 154, 156-157, 160, 162, 166
Emotions, 7, 13, 51, 140, 167-168
Empathy, 19
Enhancements, 40, 42, 44, 132, 144, 159, 172

Environmental Protection Agency (EPA), 97, 98
Ethical, 18, 22-24, 27, 30, 33, 34-35, 43, 50, 124, 152, 156, 164-166, 170-171
Ethics, 23-25, 30, 43, 62-64, 68, 79, 90, 137, 152, 158, 181
EU Commission on the Precautionary Principle, 99
EU legislation for review of consequences, 99
Europe and United States clash, 95
Europe, 91, 95, 108
Exceptional, U.S. 77
Existential risk, 2, 12, 35, 40-41, 54, 60-63, 65, 70-72, 74, 81, 90, 91, 93, 98-99, 101, 124-125, 135, 138-139, 159, 161

F

FDA, 56, 156, 163
Fear, 25-26, 51, 71, 76, 116, 120, 126, 128, 130, 138, 160
Few-shot learning, 21
Ford, Martin, 71
Fowler, Geoffery A., 20
Freeland, Chrystia, 85
Free will, 12-13, 140, 170-171
Freedom To and Freedom From, ii, 4, 57-58, 60, 87, 89, 101, 124-125, 151, 165,
Freedom From and Freedom To, 69, 100, 124, 131
Friendly, 2, 11, 40, 57, 59, 61, 64-65, 70-71, 123
Future, 1-3, 6, 11, 20, 24, 27, 29, 33-37, 40-41, 43, 47, 49, 52-54, 56, 60, 62-66, 68-69, 71-72, 75, 79, 85, 103, 108, 114, 116-117, 123, 126-127, 137-138, 141-143, 147, 150, 160-161, 169, 184

Future of Humanity Institute, 33, 68, 72
Future of Life Institute, 62, 64

G

Garnett, Kenisha, 99
Garrett, Thomas A., Federal Reserve Bank of St. Louis, 117
Generations
  Gen Z, 137
  Post-Zers, 52, 83, 136
  rebellion against boomer (s) 137
Genomics, 13, 140, 159-160
  applications of DNA, 159-161
  genetically edited babies, 162
  genetically enhancing children, 39
  genomes,159-160
  voluntary genome, 161
  voluntary modification 161
Glassco, D. Elisabeth, 81-82
Global Legal Research Directorate, 66
God of efficiency worshiped, 111-112
God of cost effectiveness worshiped, 112
Goods-producing, 106
Google Glass, 50
Government, U.S., ii, 3-4, 42, 68, 74-78, 80-85, 87, 91, 95, 98, 116, 118-119, 122, 128-130, 135-137, 147, 162

H

Hall, Edith, 137
Hanson, Robin, 69
Happiness, 2-4, 17, 42, 55, 85, 122-132, 134, 137-138, 168, 172, 111
Harari, Yuval, 70
Harris, Tom, 54
Hawking, Stephen, 72
Hinton, Geoffrey, 46-47

Hoffer, Eric, 44
HoloLens, 50
Home robot, 145
Homo Sapiens, 3, 35, 42-43
Howard, Jeremy, 151
Humans and machines, 2, 13, 139, 141-144, 157
Humanism, Secular, 35-36
Humanist Manifestos, 36
Humanity, 32, 35, 46-47, 53, 57, 61, 63-65, 67-68, 71-72, 75-76, 85, 135, 138
Humankind, ii, 1, 12, 32, 35-36, 41-42, 52, 53, 69, 71, 89, 93 101, 135-138, 146, 153, 159, 173
Humans and machines, 2, 13, 139, 142-144, 157
Humans, social animals, 128
Humanoid (s) ,1, 3-19, 21-24, 27-31, 37-39, 41-42, 45, 48, 50-52, 59-60, 62-65, 70, 72, 75-76, 85, 87, 89, 98, 100-102, 104, 110, 125-126, 135, 137, 139-141, 143, 145, 147, 148, 149-150, 152, 154, 155-159, 165, 167-168, 170-173
Husain, Amir, 71
Huxley, Aldus and Huxley, Julian, 2, 32

I

Intel RealSense technology, 145,
International Federation of Robotics (IFR), 10,15
International Policy Guidelines on AI, 66, 89
Internet of Things, 60

J

Jiankui, He, 161

Jobs, i, 7, 11-12, 65, 91, 103, 105-112, 114-115, 120-121, 138, 151

K

Kelly, Kevin, 47
Kennedy, John F., i, 134
Kristof, Nicholas, 128
Krugman, Paul, 77, 86
Kurzweil, Ray, 73, 138
Kyoto Protocol, 92

L

Lanier, Jaron, 51
Recession, 6, 79, 105, 116, 118, 120-121
Levy, David, 20, 171
Licklider, J. C. R., 140,143
Life, iii, 1-3, 7, 14-16, 23, 28, 30, 35, 37, 40-45, 48, 52-57, 60-62, 64, 67, 69, 70, 76, 84, 90, 110, 114, 122-134, 136, 138, 141, 147-160, 166-167, 169, 170-71
Lifestyles, 15, 30, 78
Lippman, Walter, 83
Love, 15-17, 19-20, 22, 29, 69, 131-132, 157

M

Machines, iii, 1-2, 13, 18, 24, 41, 47-49, 51, 54, 60-62, 69-71, 73, 109-110, 113, 134, 138-139, 141-146, 149, 150, 157, 167-170
Machine Intelligence Research Institute, 72
Machine intelligence, 70
Manufacturing, 9, 25, 105-107, 115
Maslow, Abraham, 130-131
McCarthy, John, 170-171

INDEX

Microsoft academic search list, 14
Mind, 16, 29, 38, 42, 47, 50-54, 73, 143, 151-152
Miscreants, rogues, 2, 58, 66
MIT Digital Economy Conference, 110
Mobility, 13, 139, 141,144-145, 139
Montreal Protocol, 92
Moral(s), ii, 17, 31-32, 42, 44-46, 53, 59-60, 69, 75, 80, 91, 100-101, 124, 129, 137, 163
Moratorium,161
   bans, ii, 4-5, 12, 30, 51, 69, 71, 87, 89-90, 101, 103, 126, 135, 159, 161
   regulations, ii, 3-5, 44, 80, 87, 89-91, 96, 101- 103, 125, 129, 134-135, 159, 161, 166
Mott, Mary, 166
Movement conservatism, 86-87
Muro, Mark, 8
Musk, Elon, 49, 61
Mutualism, 142-144

N

Nanotechnology, 38, 40
National Science Foundation (NSF), 69
Neural, 38-39, 46-51, 154
New Deal, 86
New generation, 83, 136, 141
Newspapers, 83-85
Niakan, Kathy, 161-162
Nietzsche, Friedrich, 17
Noonan, Peggy, 87, 121

O

O'Riordan, Tim, 91-92, 100, 104
Oligarchy, 77, 86
Olson, Parmy, 29

Organism, 3, 13, 18, 24, 34, 36, 38, 48, 94, 140, 142-144, 156-157, 159, 162-163, 166-167
Organization for Economic Co-operation and Development (OECD), 67

P

Pandemic, 20, 105, 116-121, 136
Parsons, David, 99
Paul Berg, Dr., 160-161
Pavese, Cesare, 114
Piore, Adam, 70-71
Plastic cranium human's skull, Three-D, 152
Plutocracy, ii, 4, 74-76, 77, 83, 85, 86-88, 128
Pope Francis, 27
Posthuman, 3, 32, 34, 37-39, 47, 52, 132-133, see also, posthuman condition, posthumanoid condition, posthumanity, posthumanoid
Post-Singularity, 58, 72
Power to operate, 13, 139, 148
Precautionary Principle, ii, 88, 91-96, 98-104, 124, 136
Primates, bonobos and chimpanzees, 27, 151, 164
Prosthetics, 148
Public policy, 2, 33, 43, 71, 82, 96, 123, 129
Purpose, 43, 71, 121-122, 130, 132, 137, 156, 161, 170, 172

R

Rasmussen Poll, 80
Recession, 6, 57, 80, 105, 116, 118, 120-121
Ren, Dr., 151-152

Reproduction, 13, 43-44, 139, 153, 156, 162, 165
Republican democracy, 4, 74-75, 83, 85
Republicanism, 76,
Ridley, Matt, 90-91, 101
Rights, 25-28, 63, 66-68, 75, 89-90, 92, 104, 123-126, 157, 170
Rio Conference Declaration on Environment and Development, 92
Risks, 19, 40, 47, 61-63, 66-68, 75-76, 85, 90, 95, 99-101, 123-125, 138
Robots, i, ii, 5-10, 12-15, 17-30, 37, 47-48, 51, 54, 58, 60-63, 70-71, 102, 109-113, 116, 139, 143-146, 148-150, 168, 171
Robotics, i, ii, 2, 6-7, 10-12, 15-16, 18, 23-24, 54, 60, 62, 66, 71, 102-103, 112, 145, 154
Rollbot, 149
Rosenblatt, Frank, 46
Rouse, Margaret, 19
Russell, Stuart, 70-72

S

Sachs, Jeffrey D., 85
Sagan, Carl, 1, 105
Schiefelbein, Susan, 59
Schumpeter, Joseph, 105
Sensors, 18, 65, 112, 155
Service sector, 10, 107, 111, 114-115
Singularity, 2-4, 53-54, 58, 70, 72-73, 134, 138
Simpson, Lyle, 36, 132
Skin, 22, 146-147, 166
Smith, Adam, 86
Snow, C. P., 6
Social crisis, America, 85
Source code, 1, 60
South Korean Robot Ethics Charter, 24-25, 158

Species, 1, 3, 12-13, 24, 31-32, 37, 42, 53, 135, 140-142, 163-164, 166-167, 172-173, 177
Speech, 13, 28, 49, 89, 139, 148-151, 167
Stanford Institute for Human-Centered Artificial Intelligence (HAI), 68
Steinhauer, Olen, 123
Stewart, R.B, 93, 96
Stiglitz, Joseph E., 77, 80
Stolfo, Salvatore J., 24
Supreme Court, 27, 30, 78, 95, 97, 154, 157
Symbiosis, 24, 141-144, 157

T

Tallis, Raymond, 52
Technium, the, 47
Tech Companies, 89-90, 109
Technocracy, 4, 74-75, 83, 85, 87
Technocrats, i, 4, 76, 87, 165
Technology, i, 1, 6, 14, 18-19, 29, 39, 47-48, 50, 53-54, 59-60, 65, 68-69, 71-72, 76, 81, 90-91, 94, 102, 105-106, 108-109, 111, 128, 135, 145-146, 149-151, 154, 160, 162, 170
Technomancy, 72
Tegmark, Max, 64-65, 69-70, 72
Tenner, Edward 113
Tergesen, Anne, 57
Thomas S. Foley for Public Policy and Public Service, 82
Transhumanism, 2, 3, 32-35, 37, 43
Transplants, 13, 139, 151-152, 154, 164
Triumvirate, 4, 75-76, 87, 89, 135
Trump administration failures, 119
Turing Test, 169
Turing, Alan, 169

U

UN Human Rights Council (UNHCR), 66
United Nations General Assembly endorsement, 92
United Nations Interregional Crime and Justice Research Institute (UNICRI), 66
United States, logical site for working group, 88
United States, passports, 138

V

Vanguard, 79
Vinge, Verner, 73, 138
Virtual agents, 45
Vision, eyes, 129, 139, 144-145, 146
Vision, future, 1, 13, 35, 43, 48-49, 64, 65, 81
Voeneky, Silja, 124-125
Voice commands, 8, 50-51
Voice controlled virtual assistants, 150

W

Waseda University, 10, 11
Watson, Daniel, 49, 149
Wilczek, Frank, 72
Wingspread Statement, 94
Wister, Owen, 87

Y

Yudowsky, Eliezer, 64

Acknowledgments

Thanks abound to my friend of two and a half decades, Masaru Yamada, a distinguished journalist in Japan and top in the Japan Agricultural Newspaper. It has been my privilege to have Masaru as my colleague and guide into high levels of the public and government that proved so important in my career. It has been wonderful to collaborate on three books. One, which won the Japanese Agricultural Associations' *Best Books of the Year* Award, has many hallmarks of the topics in this *AI Prevails* book.

This book would never have been possible without Bryan Tomasovich, my manager and owner of The Publishing World in Seattle. I can't praise him enough, for he has been so meticulous and caring about the book's outcome.

A special thanks goes to April Quint, my muse over a decade and a half, walking 2 ½ miles, 3 times a week. This amazing woman was continually interested and provided cogent suggestions beginning with my original project in 2006 on manifest destiny, critical thinking, and values. From there, the topic morphed into *Japan and China to the US* around 2009. In 2011, the topic changed to *Who are We? What Do We Want to Become?* April soldered on. By 2014, the title was *Deciders on Our Destiny: Humankind as We Know It*. Around then, robots joined the scene, leading to *AI Prevails* you now hold in your hands. Two years ago, April and her husband moved away. Hard to believe, April initially copy edited this book while on her way to Cyprus. Succinctly, I owe April so much. Wonderfully, she is in Seattle from May through September.

I am deeply grateful to a number of friends and professional colleagues, especially Cornell Clayton, Director of the Thomas S. Foley Institute for Public Policy and Public Service at Washington State University. In 2010, Cornell arranged for me to be Affiliate Professor and Senior Fellow. As the years have progressed, he has supported my projects, especially my *AI Prevails* book.

Thanks so much to Joanie Eppinga, who provided so much help other than her main job of heavy copyediting. Joanie, owner of Eagle Eye Editing & Writing, carried out substantive line editing and made numerous suggestions about grammar and ideas about improvement of the manuscript. She also carried out proofreading when the manuscript was in the hands of Bryan, all at a satisfying reasonable cost. She became a true friend as we talked about word usage and other things. What a gal!

My gratitude goes to my son Roderick for his great help on my writing and providing methods to improve it. My appreciation for introducing me to Sol Stein's book, *Stein on Writing*, and to *slickwrite*. See, an old dog can be taught new tricks. Significantly, he is the one who convinced me to adopt self-publication of this book.

Deep thanks to Sharleen Simpson for gifting me a love of cultural anthropology, and real experience about the culture of poverty. This is a prime reason I have been able to handle the social side in this book.

In memory of Professors Jimmy Hillman, Department Chairman of Agricultural Economics at the University of Arizona; Bob Young, my master's degree professor there; and Don Farris, my Ph.D. professor at Texas A&M University.

Last, but most important, this acknowledgment is for my beloved wife Itsuko, wife of forty years, and lights in our lives Randy (Randall), Roderick (Rod), and Roberta (July).

Author Bio

James R. Simpson is Affiliate Professor and Senior Fellow, Thomas S. Foley Institute for Public Policy and Public Service, Washington State University, Professor Emeritus, University of Florida, and Professor Emeritus, Ryukoku University, in Japan. His specialty as an international economist includes training and a career as a scientist in techniques and technologies that fit in robotics and artificial intelligence. He has focused on long-term projections of technological change as part of living and working abroad and has engaged in extensive consulting with organizations such as The World Bank. Publications include over 375 articles, monographs and software, and nine books. One of those, that has many hallmarks of the topics in this book, *Is Today's Food Situation Good for Japan? Warning from an American Researcher*, was awarded *Best Book of the Year* by the Japanese Agricultural Journalists Association.

CPSIA information can be obtained
at www.ICGtesting.com
Printed in the USA
BVHW031202270521
608293BV00009B/2309/J